# PCM and DIGITAL TRANSMISSION SYSTEMS

# TEXAS INSTRUMENTS ELECTRONICS SERIES

*Applications Laboratory Staff of*
*Texas Instruments Incorporated* ■ DIGITAL-INTEGRATED-CIRCUIT,
OPERATIONAL-AMPLIFIER, AND OPTO-
ELECTRONIC CIRCUIT DESIGN

*Applications Laboratory Staff of*
*Texas Instruments Incorporated* ■ ELECTRONIC POWER CONTROL AND DIGITAL
TECHNIQUES

*Applications Laboratory Staff of*
*Texas Instruments Incorporated* ■ MICROPROCESSORS AND
MICROCOMPUTERS AND SWITCHING
MODE POWER SUPPLIES

*Applications Laboratory Staff of*
*Texas Instruments Incorporated* ■ MOS AND SPECIAL-PURPOSE BIPOLAR INTE-
GRATED CIRCUITS AND R-F POWER TRAN-
SISTOR CIRCUIT DESIGN

*Applications Laboratory Staff of*
*Texas Instruments Incorporated* ■ POWER-TRANSISTOR AND TTL
INTEGRATED-CIRCUIT APPLICATIONS

*Bylander* ■ ELECTRONIC DISPLAYS

*Carr and Mize* ■ MOS/LSI DESIGN AND APPLICATION

*The Engineering Staff of*
*Texas Instruments Incorporated* ■ SOLID-STATE COMMUNICATIONS

*The Engineering Staff of*
*Texas Instruments Incorporated* ■ TRANSISTOR CIRCUIT DESIGN

*Härtel* ■ OPTOELECTRONICS: THEORY AND PRACTICE

*Hibberd* ■ INTEGRATED CIRCUITS

*Hibberd* ■ SOLID-STATE ELECTRONICS

*The IC Applications Staff of*
*Texas Instruments Incorporated* ■ DESIGNING WITH TTL INTEGRATED CIRCUITS

*Learning Center Staff of*
*Texas Instruments Incorporated* ■ CALCULATOR ANALYSIS FOR BUSINESS AND FINANCE

*Learning Center Staff of*
*Texas Instruments Incorporated* ■ SOURCEBOOK FOR PROGRAMMABLE CALCULATORS

*Owen* ■ PCM AND DIGITAL TRANSMISSION SYSTEMS

*Texas Instruments*
*Learning Center and*
*The Engineering Staff of*
*Texas Instruments Incorporated* ■ MICROPROCESSORS/MICROCOMPUTERS/
SYSTEM DESIGN

# PCM and DIGITAL TRANSMISSION SYSTEMS

**FRANK F. E. OWEN,** B.Sc., M.I.E.E.

Chartered Engineer

**McGRAW-HILL BOOK COMPANY**

New York   St. Louis   San Francisco   Auckland   Bogotá   Hamburg
Johannesburg   London   Madrid   Mexico   Montreal   New Delhi
Panama   Paris   São Paulo   Singapore   Sydney   Tokyo   Toronto

*Recd = 5996-1*

*Library of Congress Cataloging in Publication Data*

Owen, Frank F. E.
  PCM and digital transmission systems.

(Texas Instruments electronics series)
Bibliography: p.
  Includes index.
1.  Data transmission systems.  2.  Pulse code
modulation.  3.  Digital electronics.  I.  Title.
II.  Series.
TK5105.094       621.38′0413       81-5988
ISBN 0–07–047954–2                 AACR2

3 4 5 6 7 8 9 0   HDHD   8 9 8 7 6 5 4 3

The editors for this book were Barry Richman and Charles P.
Ray; the designer was Elliot Epstein, and the production
supervisor was Teresa F. Leaden. It was set in Times Roman
by The Kingsport Press.

ISBN 0-07-047954-2

Printed and bound by Halliday Lithograph.

# Contents

## TOMORROW'S TELECOMMUNICATION SYSTEMS

# Preface

It is indeed strange that although pulse-code modulation and digital transmission are already extremely important there have been very few textbooks written on the subject. The literature, such as it exists, has been written mainly by academics and has concentrated on the deeper theoretical aspects of the problem, rather than describing those areas that are of practical importance.

This book is intended as an overview work, and covers both the theoretical and practical problems involved. Thus, the material will be found useful to those moving into the field for the first time, as well as engineers engaged in the development, operation, and planning of the described equipments. The latter group should find the text most suitable as a reference work, and are directed to the appropriate technical papers should more details be required.

Although the field of digital telecommunications is clearly the subject of this book it should be pointed out that the described techniques will find applications elsewhere. For example, recent developments in digital sound recording, digital radio transmission, and the digitization of video signals have all encountered the same engineering limitations outlined within the text. Consequently engineers practicing in such related areas are also likely to find this book useful.

The book is organized into five parts as follows:

Part 1   Introduction (A general overview of telecommunication practices)

Part 2   Pulse-Code Modulation

Part 3   Multiplexing

Part 4   Transmission

Part 5   Tomorrow's Telecommunication Systems

Each section is reasonably self-sufficient, and the reader will find no difficulty in changing the order in which they are read.

Each of the individual topics begins with a general overview of the problem. As the text proceeds the descriptions become more technical. Thus, for those readers who seek a basic outline of the techniques involved it will be sufficient to concentrate their reading on the earlier description within each section.

Although this book has been written primarily with the professional engineer in mind, college students will also find it useful for both home and course study. It is recommended that college professors follow the basic arrangement of the book when planning their lecture series. Each chapter represents enough material to satisfy approximately a one-hour lecture. Most students will find Chapters 4 (Quantization), 7 (Asynchronous Time Division Multiplexing), 10 (Timing Extraction and Jitter), and 11 (Equalization) particularly arduous due to the bulk of information that needs to be digested, rather than its complexity. This should be taken into account when planning the timing allocation of individual lectures.

# ACKNOWLEDGMENTS

I am indebted to the understanding and support of my wife Dilek, and our children, Sidika and Gustav, during the writing of this book. I really doubt that the manuscript would ever have been completed had my family been unprepared to make the necessary sacrifices in giving up our leisure time over what seemed a very long period.

It is doubtful whether a book of this size, covering so many technical fields, could be uniquely authored by a single individual. I have received considerable help from a number of sources, notwithstanding those I have never met and know only via their published work. I am indebted to my old colleagues within the various International Telephone and Telegraph company centers in Europe, in both the Universidale Estadual de Campinas, São Paulo, Brazil, and the Centro de Pesquisa e Desenvolvimento Telebras, Campinas, Brazil, as well as to the engineers of Texas Instruments.

I must pay particular tribute to Professor B. P. Lathi, who pushed me into writing the book in the first place and gave me invaluable encouragement and technical assistance during the earlier chapters. Professor R. Scarabucci, also of Campinas University Brazil, gave considerable encouragement, and theoretical assistance in the area of companding analysis.

I would like to thank Professor Cattermole (Essex University), and Ton Tanke (Texas Instruments, Germany) for their valuable inputs related to the preliminary drafts of the manuscript.

Finally, I would like to express my thanks to Julia at Campinas University and to my secretary, Elisabeth Petitjean, who typed the manuscript.

Frank F. E. Owen

# Abbreviations for Units

| | |
|---|---|
| A | amperes |
| Bd | baud |
| bit/s | binary digits per second |
| cm | centimeters |
| dB | decibels |
| dB/km | decibels per kilometer |
| FITS | failure in tens of seconds |
| Gbit/s | gigabits (1,073,741,824 binary digits) per second |
| GHz | gigahertz |
| Hz | hertz |
| h | hours |
| kA | kiloamps |
| kbit/s | kilobits (1024 binary digits) per second |
| kHz | kilohertz |
| km | kilometers |
| Mbit/s | megabits (1,048,576 binary digits) per second |
| MHz | megahertz |
| m | meters |
| $\mu$s | microseconds |
| mA | milliamperes |
| mm | millimeters |
| ms | milliseconds |
| mW | milliwatts |
| min | minutes |
| nA | nanoamperes |
| nF | nanofarads |
| nm | nanometers |
| ns | nanoseconds |
| ns/km | nanoseconds per kilometer |
| $\Omega$ | ohms |
| pF | picofarads |
| ppm | parts per million |
| s | seconds |
| V | volts |
| W | watts |

# PCM and DIGITAL TRANSMISSION SYSTEMS

# INTRODUCTION

# 1

# History of PCM
# and Its Application

This book aims to satisfy the obvious need for a comprehensive description of the digital communication systems that exist today, and those that are envisaged for tomorrow. The earlier descriptions are nontechnical and in consequence are likely to satisfy those persons who seek a basic outline of the techniques involved. The later descriptions, however, are very technical and should provide a first reference for equipment designers and students of telecommunications.

At the outset it is important to note that the digital techniques described here have found many diverse fields of application. Possible examples include outer space probes, high fidelity music, television recording, voice-frequency data transmission, and so on. The list is endless, however the most important and widest usage to date for the digital approach is in telephony. Consequently we shall study the telephonic application in detail, and note that the same techniques are equally applicable elsewhere.

The public telephone network, which it should be remembered is based mostly upon outdated technology, is currently being pressed for a tremendous expansion both in capacity and in the facilities provided. The number of telephone subscribers has, in most countries, doubled during the last 10 years, while utilization has grown at a much greater rate. Studio quality music and color television links are now accommodated within the network, while new services such as Confra-Vision[1] and facsimile[2] are being considered. It is, however, in the area of data transmission that the greatest expansion has taken, and indeed is taking, place.

Within this chapter we shall concern ourselves with brief details of how and why we may decide to transmit signals digitally; noting that most naturally occurring messages have an analog character. Chapter 2 aims to briefly summarize the history of telephony development, and how a telephone network is arranged today. A rudimentary understanding of this will help us later in the book, when we consider specific transmission problems.

The technical material has been divided as far as possible into three separate areas, which are presented in the following order:

---

[1] A system for connecting two centers by television plus sound so that a conference may take place.

[2] A system for transmitting over a period of time a photographic image, using a very low bandwidth link. The image should contain sufficient detail so that typed script may be resolved.

3

   1. Conversion of an analog signal into a digital one
   2. Combining several digitized signals together, that is, *multiplexing*
   3. Transmission of digital signals

It is true that most information signals that are to be transmitted are analog. A typical example may be the signals produced by the microphone within the subscriber's handset. These analog signals need to be converted to the digital format, and the most widely used technique is known as PCM, an abbreviation for *pulse code modulation.*

PCM was patented in 1939 by Sir Alec Reeves, who was at the time an engineer of the International Telephone and Telegraph Company (ITT) laboratories in France. He proposed a technique which involved sampling the information signal at regular time intervals, and coding the measured amplitude value into a sequence of pulses. The pulses became the transmitted signal and they conveyed an impression of the message as a series of binary numbers. At the receiver the binary numbers were used to reconstruct the original analog signal in a similar way to plotting a graph using known plots to indicate the form of the required curve.

PCM is dependent on three separate operations, *sampling, quantizing,* and *coding.* Many different schemes for performing these three functions have evolved during recent years, and in Chaps. 3, 4, and 5 of this book we shall describe the main ones. In these chapters we shall see how a speech channel of telephone quality may be conveyed as a series of amplitude values, each value being represented, that is, coded, as a sequence of 8 binary digits. Furthermore, we shall prove that a minimum theoretical sampling frequency of order 6.8 kilohertz (kHz) is required to convey a voice channel occupying the range 300 Hz to 3.4 kHz. Practical equipments, however, normally use a sampling rate of 8 kHz, and if 8 digits per sample value are used, the voice channel becomes represented by a stream of pulses with a repetition rate of 64 kHz. Figure 1-1 illustrates the sampling, quantizing, and coding processes.

Reexamination of our simple example shows us that the speech signal of maximum frequency 3.4 kHz has been represented by a signal of frequency 64 kHz. However, if only 4 digits per sample value had been used, the quality of transmission would drop, and the repetition rate of the pulses would be reduced to 32 kHz. Thus the quality of transmission is dependent on the pulse repetition rate, and for digital communication systems these two variables may be interchanged most efficiently.

Digital transmission provides a powerful method for overcoming noisy environments. Noise can be introduced into a transmission path in many different ways; perhaps via a nearby lightning strike, the sparking of a car ignition system, or the thermal low-level noise within the communication equipment itself. It is the relationship of the true signal to the noise signal, known as the signal-to-noise ratio, which is of most interest to the communication engineer. Basically, if the signal is very large compared to the noise level, then a perfect message can take place; however, this is not always the case. For example, the signal received from a satellite, located in far outer space, is very weak and is at a level only slightly above that of the noise. Alternative examples may be found within terrestrial systems where, although the message signal is strong, so is the noise power.

If we consider binary transmission, the complete information about a particular message will always be obtained by simply detecting the presence or absence of the pulse. By comparison, most other forms of transmission systems convey the message information using the shape, or level of the transmitted signal; parameters that are

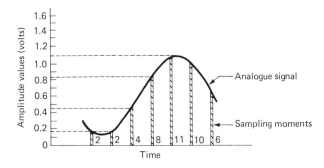

| Amplitude value | Binary coded equivalent | Pulse code modulated signal |
|---|---|---|
| 1 | 0 0 0 0 | |
| 2 | 0 0 0 1 | |
| 3 | 0 0 1 0 | |
| 4 | 0 0 1 1 | |
| 5 | 0 1 0 0 | |
| 6 | 0 1 0 1 | |
| 7 | 0 1 1 0 | |
| 8 | 0 1 1 1 | |
| 9 | 1 0 0 0 | |
| 10 | 1 0 0 1 | |
| 11 | 1 0 1 0 | |
| 12 | 1 0 1 1 | |
| 13 | 1 1 0 0 | |
| 14 | 1 1 0 1 | |
| 15 | 1 1 1 0 | |
| 16 | 1 1 1 1 | |

If the analogue signal shown above is "sampled", and then "coded" using the table, the transmitted pulse code modulated signal becomes:

Decimal values:     2  ,   2  ,  4  ,  8  ,  11  ,  10  ,  6

Binary values:     0 0 0 1 , 0 0 0 1 , 0 0 1 1 , 0 1 1 1 , 1 0 1 0 , 1 0 0 1 , 0 1 0 1 ,

PCM signal:

**Fig. 1-1**  The sampling and coding processes, and the resultant PCM signal.

most easily affected by the noise and attenuation introduced by the transmission path.  Consequently there is an inherent advantage for overcoming noisy environments by choosing digital transmission.

In extreme cases, when the signal-to-noise ratio is particularly poor, we may use redundancy techniques.  This has been successfully used for video, and data transmission on the Apollo space flights.

By using redundancy the message may be repeated many times, or otherwise extended, and single digit errors that occur within the detected signals are ignored.  For example, we have seen that an amplitude value can be coded as a sequence of 8 pulses, perhaps 10000000, where 1 represents the presence of a pulse, and 0 its absence.  The effect of noise will be to degrade the purity of reception, thus the

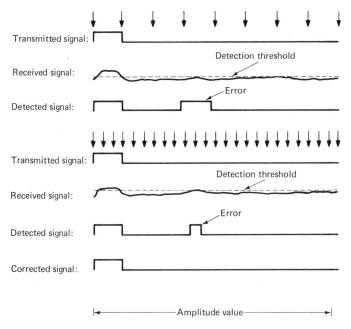

**Fig. 1-2**   Signal processing by use of redundancy.

sequence 10000000 may be misinterpreted as 10010000, which in fact represents a completely different amplitude value.   Here an error has been introduced at the fourth digit location, due to the receiver being confused between the weak true signal, and the surrounding noise.

If now the sequence of 8 digits is increased to 24 digits such that the above message is transmitted as 111, 000, 000, 000, 000, 000, 000, 000, then single digit errors can easily be detected[3] and ignored by simple digital processing, as shown in Fig. 1-2. In this example we have increased the number of transmitted digits, without increasing the quantity of information communicated, in order to combat noise.   In practice, one is likely to adopt a more complicated error correcting system than identified here, but the point is made that digital systems lend themselves most easily to signal correction within the receiver.

So far in this discussion we have assumed that each voice channel has a separate *coder,* the unit that converts sampled amplitude values to a set of pulses; and *decoder,* the unit that performs the reverse operation.   This need not be so, and systems are in operation where a single *codec*[4] (i.e., coder, and its associated decoder) is shared between 24, 30, or even 120 separate channels.   A high-speed electronic switch is used to present the analog information signal of each channel, taken in turn to the codec.   The codec is then arranged to sequentially sample the amplitude value, and code this value into the 8-digit sequence identified earlier.   Thus the output to the codec may be seen as a sequence of 8 pulses relating to channel 1, then channel 2, and so on.   This unit is called a *time division multiplexer* (TDM), and is illustrated

---

[3] Note: If the receiver detects 010, or 101, an error has occurred.   This can be: (a) detected as an error, for error-rate measurement; and (b) corrected to 000, or 111 respectively.   (See Fig. 1-2.)

[4] See Chap. 12 for switching systems that employ single channel codecs, rather than the shared codec approach described here.

in Fig. 1-3. The multiplexing principle that is used is known as *word interleaving,* since the words, or 8-digit sequences, are interleaved in time.

At the receive terminal a demultiplexer is arranged to separate the 8-digit sequences into the appropriate channels. The reader may ask, how does the demultiplexer know which group of 8 digits relates to channel 1, 2, and so on? Clearly this is important! The problem is easily overcome by specifying a *frame format,* where at the start of each frame a unique sequence of pulses called the *frame code,* or *synchronization word,* is placed so as to identify the start of the frame. The circuit of the demultiplexer is arranged to detect the synchronization word, and thereby it knows that the next group of 8 digits corresponds to channel 1. The synchronization word reoccurs once again after the last channel has been received.

The time division multiplexer forms the basis of any digital communication network, and we shall discuss it in detail later in this book. Fortunately, it is a reasonably cheap unit since its implementation is based on readily available mass-produced digital

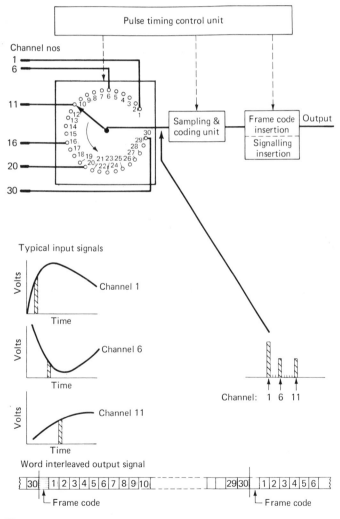

**Fig. 1-3** The function of the time division multiplexer (TDM).

integrated circuits.  Consequently its use is economically viable over even short distances.

Let us consider the economic arguments for a moment.  In telephony there is a requirement for many links of distance 20 to 50 km.  If the multiplexing equipment is very expensive, then it is clearly more economic to use many separate cables each carrying a single speech channel.  This has happened until recently, since the previously available frequency division multiplexers (FDM) were extremely expensive and became economic only when used over very long distances.  Now that it is no longer true that multiplexing need be expensive, TDM equipment is currently being introduced on even very short routes.

The widespread introduction of digital transmission equipment within the telephone networks has not been rapid since the original French patent was issued in 1939.  Most of the early work was carried out in the United States, where the Bell Laboratories produced several experimental systems based on vacuum coding tubes.  These were bulky and expensive to produce.

The availability of cheap transistors changed the economics of PCM coders and TDM multiplexers considerably.  As a result, in 1962 full-scale production of the Bell T1 transmission system began at American Telephone and Telegraph Corporation.  Since that date the introduction of cheap integrated circuits (IC's) has ensured a firm place for digital equipments within the telephone network.

Today the usage of PCM systems within the United States is extensive, as it is in the United Kingdom and Japan.  In addition, many other countries have made major investments in digital equipment for their national telephone networks.  The list now includes Belgium, Italy, Scandinavia, and South Africa.

It is clear that in the near future, digital signals will be used on many international routes.  For this reason the International Telegraph and Telephone Consultative Committee[5] (CCITT) has during recent years agreed on a transmission hierarchy on which many national PCM networks are now based.

Two main hierarchies exist within the world today (see Fig. 1-4), those based on a time division multiplexer that word interleaves 30 separate speech channels (European, African, South American scheme), and those that word interleave 24 separate speech channels (North American, Canadian, Japanese, and the earlier British system).  The functions performed by these equipments are identical, only the frequencies involved are different due to the greater bandwidth requirement for accommodating 30 channels compared to 24.  We shall use the 30-channel basic hierarchy as an example in this book, since it has the widest global coverage.  However, where appropriate, the corresponding 24-channel equipments will be identified in order that readers from countries served by these types may relate to them.

The *binary digit information rate,* or bit rate, at the output of the 30-channel multiplex is 2.048 megabits per second (Mbit/s).  This bit rate is obtained by interleaving 32 words, or time slots, before the cycle is repeated.  Thirty of these words contain the amplitude values from the 30-channel voice signals, while the remaining two time slots contain digits that are used for the synchronization of the demultiplexer to the multiplexer, and signaling information.  Digital communication engineers have found many different ways of naming this equipment, which can be confusing at first.  The terms *30-channel multiplex, 2 Mbit/s multiplex, 32 time-slot muliplex,* and *primary multiplex* are all in common use, but they all refer to the same equipment.

---

[5] This committee, composed of representatives from the member countries, formulates operating standards for telegraph and telephone equipment, which is used on, or influences, international routes.

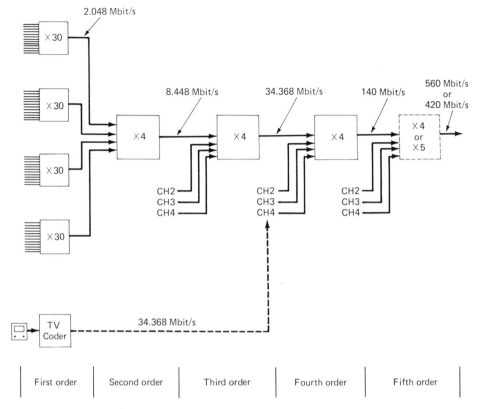

**Fig. 1-4a**   The PCM hierarchy as is used in: Europe, Africa, Australasia, and South America.

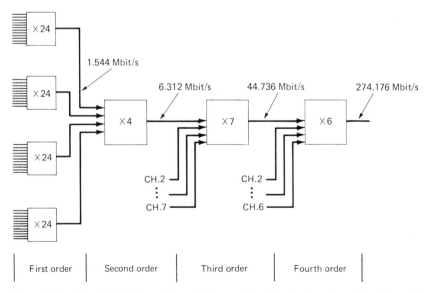

**Fig. 1-4b**   The PCM hierarchy as it is used in North America, Canada, and Japan.

The second-order equipment identified in Fig. 1-4*a* contains a multiplex that combines four signals each of bit rate 2.048 Mbit/s into a signal of rate 8.448 Mbit/s. This multiplex contains no encoder and performs the required operation by simply bit interleaving the four input signals.   The resultant 8.448 Mbit/s output stream has consecutive digits from different channels in the form: bit A channel 1, bit A channel 2, bit A channel 3, bit A channel 4, bit B channel 1, etc.   At the demultiplexer the original channel signals are reconstructed and the word-interleaved structure is seen again.   Second-order multiplexers of this type are currently in service within Italy, Scandinavia, Belgium, and South Africa.

Already equipments are in operation at the third and fourth orders of the hierarchy, using bit rates of 34.368 Mbit/s and 139.264 Mbit/s respectively.   The fifth order, at the time of writing, still lacks international agreement, but is likely to have a bit rate of 560 Mbit/s (see Fig. 1-4*a*).

The equivalent bit rates for the 24-channel hierarchical structure (see Fig. 1-4*b*) as used in North America,[6] Canada, and Japan are:

First order, Bell T1 system =    1.544 Mbit/s
Second order, Bell T2 system =    6.312 Mbit/s
Third order, Bell T3 system =   44.736 Mbit/s
Fourth order, Bell T4 system = 274.176 Mbit/s

The earlier British 24-channel system, which continues to be used nationally, has now been superseded by the European 30-channel standard.   The British equipment had a different structure from that used in the Bell system and operated at the following transmission rates:

First order =    1.536 Mbit/s
Second order =    6.336 Mbit/s
Third order = 120.0    Mbit/s

Judging by the number of PCM equipments that exist worldwide, PCM in particular, and digital communications in general is already an important telecommunication method.   We have noted that since the introduction of semiconductor devices multiplex equipment costs have reduced, and that digital transmission techniques are able to combat poor transmission media.   These, however, are only a few of the many advantages afforded by a network based on the transmission of digital signals.

It is true of any communication system that the shape and amplitude of the transmitted signal will be continuously degraded by the introduction of noise, and the attenuation along the transmission path.   The format of a digital signal is regular and we need only to detect the presence or absence of a pulse at each bit position.   The shape of the pulse is unimportant.   Consequently, provided the degradations are kept within certain limits the original signal may be interpreted from the received, imperfect one.   This process is known as *regeneration,* and is a very important property of PCM communication.

Imagine that a digital signal is to be conveyed via a cable over a very long distance. After the signal has traveled a few kilometers it becomes attenuated, and loses its shape as shown in Fig. 1-5.   The signal may now be amplified and regenerated such that a signal identical to the original is passed along the cable for further transmission.

---

[6] Contribution 159 to CCITT Special Study Group D, International Telecommunications Union, Geneva, Switzerland.

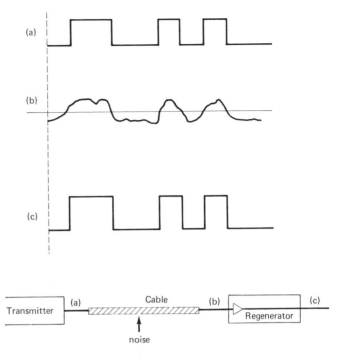

**Fig. 1-5** The regeneration process: (a) the transmitted code; (b) the received code (attenuated, and affected by noise); (c) the regenerated code.

After a few more kilometers the process is repeated, since *regenerative repeaters,* (the unit that amplifies and regenerates) are spaced at equal distances apart along the whole length of the cable. As a result, the quality of the communication becomes independent of both the transmission distance, and the topology of the route.

In this way the signals produced by the 30-channel multiplex may be conveyed at a rate of 2.048 Mbit/s on cables that were originally intended for analog voice signals of maximum frequency 3.4 kHz. Certainly transmission at such a high bit rate would not be possible over any reasonable distance [greater than about 3 kilometers (km)] if it were not for the regenerative property.

Fortunately, convenient access points for inserting the regenerative repeaters within the voice-frequency cables already exist, at equally spaced distances of about 2 km along the cable. These access points were used to house the loading coils that were placed across the cable pairs when voice-frequency traffic was transmitted. In a PCM system the coils are modified and are replaced by a regenerative repeater. Consequently, the introduction of a PCM system is achievable with only a minimal disturbance to the existing network (see Chap. 2).

Thus digital transmission techniques enable the channel capacity of *existing* cables to be increased from 1 to 30. This feature is very attractive to telephone operating companies, since it permits existing routes to be gradually increased in channel capacity, as and when required. Furthermore the need for expensive and time-consuming cable laying operations may be avoided.

A network based on the transmission of digital signals is an extremely flexible one. Once a telephone signal or a color television signal has been converted to the digital format they are completely compatible with other digital signals. Thus digitized television, music, and telephone signals may be multiplexed together and transmitted as if they were one signal. Data and telex signals, which naturally have a digital format, may be included within the composite signal also. Consequently, if the cables and repeaters are designed to always accommodate digital signals, many different types of information may be communicated via the link by changing the configuration at the terminal only. Once again, cost analysis of competitive systems that also offer flexibility have been unfavorable compared to the digital approach.

We shall see later in this book that digital transmission lends itself most readily to the new types of media envisaged for the networks of tomorrow. Currently, cable, radio, and satellite links are in use, while in the very near future optical communication, using lasers and long-haul waveguide systems are envisaged.

PCM signals provided by time division multiplexers have a regular and predictable format. We know for example, that a pulse sequence related to channel 3 always occurs after that related to channel 2 and so on. Additionally, we know that we can guard against single digit errors as described earlier. If this is possible we can also detect the occurrence of single bit errors. Thus, due to the expectation of a regular signal and the ability to detect single errors we have the opportunity of providing powerful monitoring systems that can not only detect the failure of a transmission link, but monitor the quality of transmission on that link also. Supervision techniques will be described in Chap. 8.

Data and telex information messages already possess a digital format and are consequently most easily transmitted alongside PCM signals. Usually such signals have a low bit rate (telex: 50 binary digits per second (bit/s), or 60 bit/s; data: 600, 2400, or 9600 bit/s) and it is possible to accommodate many of them within a single PCM channel of transmission capacity 64 kilobits per second (kbit/s). The possibility of providing national and international data networks using cheap, readily available 64 kbit/s transmission paths is receiving careful study in many parts of the world.

The signaling information originating from the dial at the subscriber's premises also has a digital structure. In a conventional system the dial pulses are used to route the call via various transmission paths until the required destination is reached. The process is known as *switching*.

To date in the history of telephony there have been two distinct disciplines: transmission, and switching. An engineer worked either in the transmission area, or in that of switching, but not both. Due to the introduction of PCM and its digital structure this situation is changing. Digital switching[7] is becoming a reality.

*Digital switching* is the term used to express the technique of switching where the PCM information signal and the associated signaling pulses use the same multiplexing structure. In this way, the switching circuit "knows" which sequence of pulses within the PCM information stream relates to channel 1, channel 2, and so on. Thus, the switching unit is able to detect the signaling pulses, and under their command arrange for each of the multiplexed channels to be switched to the desired destinations.

The first major exchange incorporating digital switching was installed in 1969 by the British Post Office, at Moorgate, using equipment designed by Standard Telephone and Cables (STC), the British subsidiary of ITT. Since that date several other centers

---

[7] I use this term here although the technique is sometimes referred to as *integrated switching*. The French use the term *numerical switching*.

have been brought into operation with considerable success, and it appears likely that these exchanges will point the way to those of the future.    The benefits offered by digital switching systems, simply stated, are these:

1. There is a significant cost reduction, when compared to conventional exchanges.
2. The transmission and switching units are similar in operation.    Thus, the components and power supplies that are used may be almost identical.    The temperature, humidity, and other environmental considerations may also be equated.
3. No coding process is required, since compatibility between the information and switching signals already exists.    They are both digital.

It should be pointed out that digital switching schemes are not limited to telephone applications alone.    Exchanges designed to accept data and telex signals exclusively are already in operation.

In this introduction I have concentrated on describing the most normal usages of PCM systems.    However, for completeness it is necessary to mention briefly another area of communication where PCM is employed.

Conversations requiring high security, such as in military applications, are conveniently handled by PCM equipments.    The analog-to-digital coding may be achieved in a number of ways; and the digital signal may be processed by digital scrambling techniques such that the resultant transmission is unintelligible to unauthorized users. The size, weight, and ruggedness of PCM and TDM equipments compared to alternative, competitive systems favor the digital approach very strongly.    Moreover, the maintenance interval of digital equipments is long.    This is a convenient point to summarize the reasons why digital transmission systems have become so prominent.

## SUMMARY

1. NETWORK COST

   The capital equipment investment that is involved in equipping the telephone network of a nation is tremendous by any standards.    The sums of money involved are very much larger than those contemplated by car manufacturers, oil refining companies, and the like.    Consequently, the cost of new equipment is always a major consideration to any telephone operating company.    PCM/ TDM equipment is cheap enough to be used on even short routes, and this approach provides an overall reduction in total network costs.

2. GRADUAL TRANSFER FROM THE EXISTING NETWORK TO ONE BASED ON PCM/TDM IS POSSIBLE

   PCM/TDM may be introduced gradually at the local transmission level within the telephone network.    With the PCM 30 system the channel capacity of *existing* cables is increased from 1 channel to 30 channels for each voice-frequency cable pair that is converted.    The disturbance to the existing network is minimal.

3. NETWORK FLEXIBILITY

   Once a voice channel has been converted to a digitally equivalent signal, it has a form identical to other digitally coded signals.    Encoders already exist for converting studio quality music, color television, facsimile, video-phone, etc., to the digital form.    The digital signals may be multiplexed, switched, transmitted, and their quality monitored by standardized equipment.    Additionally, any of the above digital signals may be multiplexed together with signals that are already digital by nature (e.g., data, telex, and signaling information).

### 4. NETWORK SUPERVISION
The adopted format of a PCM/TDM signal is regular and predictable. Consequently, any departure from the determined structure is immediately recognized as a fault. Information about the quality of the transmission link may be obtained by examining the number of occasional departures from the normal sequences that occur in a given time interval (digital error rate). This provides an effective method for predetermining an eventual link failure.

### 5. NETWORK MAINTENANCE
The service interval of PCM transmission equipment is long.

### 6. REGENERATION
The ability to recover the originally transmitted digital signal exactly from a degenerated received signal is extremely important. As a consequence, the quality of the communication becomes independent of both the transmission distance and the topology of the route.

### 7. SIGNAL PROCESSING
Extremely poor transmission media may be overcome by using redundant error correcting digital coding, and processing the received signal. Additionally, by using a digital scrambler the transmission may be rendered unintelligible to unauthorized users, thus providing information security.

### 8. SUITABILITY TO DIFFERENT TYPES OF TRANSMISSION MEDIA
PCM signals have been successfully transmitted via cables, radio link, satellite link, optical waveguide (multimode, and single mode), and long-haul waveguide.

If there is one reason why there is so much interest in digital communications, it is probably its cheapness compared to competitive systems. However, the truth is that it is the combined weight of the arguments favoring the digital approach that have led to the rapid introduction of PCM in so many countries already.

Within this decade there are likely to be telephone networks based on the principle of PCM in the following countries: United States, Canada, Mexico, Brazil, Australia, South Africa, Britain, Belgium, France, Italy, Norway, Sweden, Denmark, Russia, Spain, West Germany, Austria, Switzerland, and Japan. This is surely proof that PCM will be a major force in providing world telecommunications. Perhaps it already is.

# 2

# Transmission Requirements and Methods

Any student of telecommunications will notice that the most important usage to date for the digital approach is in telephony.  For this reason it is useful to briefly describe standard telephonic practices, and how a "typical" telephone network is arranged.  Hopefully this will promote a better understanding of the various practical transmission problems that exist, and will enable us to appreciate the significance of the digital approach.

## 2.1  THE EVOLUTION OF THE TELEPHONE NETWORK

A national telephone network is not the product of a well-conceived master plan. It evolves a little at a time and lags behind the latest available technology.  Perhaps the best example of this is to be found in the development of switching technology, the history of which we describe here.

Just over 100 years ago, on March 10, 1876, the first successful transmission of intelligible speech was achieved by Alexander Graham Bell.  This was the culmination of a year of intensive experiments carried out by Bell and his assistant, Thomas Watson, working in an attic in Boston.

A further year of continous experimenting was required to develop the invention into a practical form.  During this time, successful tests were performed over lines of increasing length and public demonstrations were given.  Bell and his associates then commenced the manufacture and leasing of telephone sets, and a public telephone service was born.  From this beginning has developed the present worldwide telecommunication network with over 350 million telephones.

The interconnection of subscribers in the earliest telephone networks was performed manually.  The telephone operator, whose attention was drawn by an electric bell, physically wired the requested connection.  The system has been improved during the years, but the same basic concept is still in use within small villages, private telephone networks, and in areas where labor costs are low, such as the developing nations.

Manually operated telephone exchanges have high operating costs and are slow. However, strangely enough, cost and speed were not the main considerations that promoted the introduction of the world's first automatic telephone exchange.

In a small American town there existed two funeral directors.  One, whose name was Strowger, became anxious when he noticed that his clientele diminished to almost zero, while his competitor had an astounding increase in business. Strowger investigated, and discovered that his competitor's wife had recently begun employment as the town's telephonist. Clearly, it was she who was informing the competitor about "new business" as it happened! Strowger decided to devote his energies to designing an automatic exchange in order that his competitor's wife might be replaced. He succeeded and produced a device known as a *Strowger switch,* or *uniselector,* which is the basis of most automatic telephone exchanges in existence today.

The Strowger switch is an electromechanical device which responds to the pulses provided by the dial on the subscriber's telephone set. Each pulse rotates the switch by one position, where a new set of contacts exists. By carefully arranging the Strowger switches, it is possible to provide a system that routes a transmission path through the exchange under the direction of the dial pulses. This was Strowger's real achievement.

Electromechanical switching systems are expensive, difficult and costly to maintain, slow in operation, and large. However, a large investment already exists in this outdated technology, and the more modern switching systems that have been and are being developed must be compatible.

In recent years many of the problems associated with an electromechanical system have been avoided by using a reed relay as the switching element. This device, which is illustrated in Fig. 2-1, is glass encapsulated and is surrounded by a solenoid. When a current is passed through the solenoid, the magnetic field that is produced causes the normally open reeds to close. (In certain types the reeds continue to remain closed, due to magnetic latching, even when the solenoid is deenergized. The reeds are opened by passing a current in the reverse direction through the solenoid.) The interpretation of dialing pulses into switching commands may be performed electronically; a faster, cheaper solution than the mechanical equivalent. Moreover, since the switching action is performed in a vacuum, maintenance is not required and reliability is increased.

An interim step in the development of switching systems was the use of semiconductor crosspoints as the switching element. The adoption of such devices heralded the era of true fully electronic switching, and gave the usual associated benefits of improved reliability, low cost, reduced maintenance, and so on. The so-called crosspoint is in fact an array of thyristors and transistors arranged in a matrix format (4 × 4 is typical). The switching action is performed by causing a low-resistance path to be formed between points within the matrix, under the control of an appropriate steering signal.

To date crosspoint switching has been employed mainly in small-scale switching systems of order less than 50 lines, rather than in larger machines (e.g., 2000 lines), where PCM or reed-relay switching has traditionally appeared to be more economical. Thus the crosspoint has been widely used within local concentrators feeding major

**Fig. 2-1** The reed relay: (a) contacts open—no magnetic field; (b) contacts closed—magnetic field.

switching machines, and in small private automatic branch exchange (PABX) equipments.

The switching technologies identified above may be classified as *space division switches.* The latest development in fully electronic switching has been the arrival of time division multiplexed switches based on the same digital PCM techniques which are the subject of this book.

It is pertinent to point out that PCM messages are not the only signal that may be switched in the time domain. Pulse amplitude modulated (PAM), and delta modulated (DM) signals (see Chap. 3) have also been successfully switched in equipments of recent design. However, due to the availability of cheap digital integrated circuits, PCM is now becoming more frequently used in new switching equipment designs.

A PCM based TDM switching machine can only switch digital signals, and analog voice messages must be converted to the digital format prior to switching. The conversion is achieved using standardized PCM coding rules in an identical manner

**Table 2-1a  Examples of Various Switching Equipments in Europe**

| Country | Use | Electromechanical | Half-electronic reed relay | Fully electronic crosspoint | Fully electronic PCM TDM |
|---------|-----|-------------------|----------------------------|-----------------------------|--------------------------|
| Sweden | PTT | | AXE system[e] | | Digital AXE[e] |
| | PABX | | | | |
| France | PTT | Pentaconta[j] CP400[a] | Metaconta[d] | | E10[a] MT20[j] |
| | PABX | | | P30, P40[j] | |
| Germany | PTT | Strowger[g] | EWS[g] | | System 12[d] EWS-D[g] |
| | PABX | | EMS[g] | System 40/30[h] | |
| Italy | PTT | Strowger | AXE[b] Metaconta[d] | | System 12[d] DTNI[i] |
| United Kingdom | PTT | Cross bar[c,d,f] Strowger[c,d,f] | TXE 2/4/4a[c,d,f] | | System X[c,d,f] |
| | PABX | | | | CDSS1[c,f] PDX[f] SL1[c] |
| Belgium | PTT | Pentaconta[k] | Metaconta[k] | | System 12[k] |

[a] Compagnie Industrielle des Telephones, France.

[b] Fabb. Appar. Telef. E Mater. Elet. Brev Ericsson, Italy.

[c] General Electric Co., England.

[d] Standard Telephone and Cables (International Telephone and Telegraph), England.

[e] L. M. Ericsson, Sweden.

[f] Plessey Telecommunications Ltd., England.

[g] Siemens AG, Germany.

[h] Telefonbau U Normalzeit Lehner and Co., Germany.

[i] Telettra SPA, Italy.

[j] Thomson Brandt SA, France.

[k] Bell Telephone Manufacturing Co., S.A. (International Telephone and Telegraph), Belgium.

**Table 2-1b    Examples of Various Switching Equipments in the United States, Canada, and Japan**

| Country | Use | Electromechanical | Half-electronic reed relay | Fully electronic crosspoint | Fully electronic PCM TDM |
|---|---|---|---|---|---|
| United States | PTT | C1 EAX[c] | ESS no. 1–3[a] 1 EAX[c] 2 EAX[c] | ET SS[c] | ESS no. 4[a] |
| Canada | | As above | As above | As above | DMS 1[f] DMS 10[f] DMS 100[f] |
| Japan | | C23SE[d] | ND10[e] ND20[e] FETEX 100L[b] | | HDX10[d] NEAX[e] |

[a] Bell Telephone Laboratories, United States.
[b] Fujitsu Limited, Japan.
[c] General Telephone and Electronics, United States.
[d] Hitachi Electronics Co., Japan.
[e] Nippon Electric Co., Japan.
[f] Northern Telecommunications Limited, Canada.
SOURCE: From G. A. Langley, "Beginnings: The Global Transition to Digital Switching Is Underway," *Telephony,* vol. 195, 1978, pp. 104–131.

to that used by a transmission multiplexer. Thus, the need to multiplex for both transmission and switching can be realized within the same equipment, giving an obvious economic advantage.

It is beyond the scope of this book to describe the different PCM switching philosophies that already exist, and those that are in development. However, a few basic comments may be made:

1. Early systems such as the E10 system employ analog crosspoint concentrators followed by PCM time division multiplexed switching. This French system is claimed to be the world's first commercial application of PCM switching within a central office.
2. The American systems to date have been aimed at very high-capacity equipments. The most famous example is Electric, which was first installed in Chicago and introduced into service in January 1977. It is one of the largest switching systems in the world and is capable of handling 107,520 lines.
3. More recent designs such as the British Post Office System X, or the ITT System 12 equipments employ PCM concentrators. These equipments illustrate that digital switching of even small numbers of analog voice signals can be realized economically. This is a breakthrough! We shall return to discussing such equipments in Chap. 12, where we consider the impact of low-cost single-channel PCM codec integrated circuits.

By means of a summary a few examples of various equipments currently in service are shown in Table 2-1, together with the switching technology employed.

For the remaining sections of this book we shall consider the transmission aspects of PCM, and the associated digital techniques. However, it is pertinent to stress that the same practices are equally applicable to digital switching.

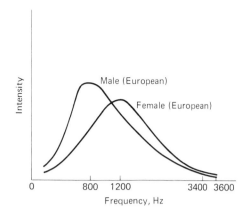

**Fig. 2-2**   The spectrum of human speech.

## 2.2   TELEPHONY TRANSMISSION QUALITY

The quality of communication offered to the telephone user may be expressed in terms of frequency response, harmonic distortion, signal-to-noise ratio, and so on. However, establishing values for these parameters is difficult since the term *quality* is imprecise and subjective.   It is clear that the listener must be able to recognize the meaning of the spoken word, the nuances of speech, and indeed the identity of the speaker.   Furthermore, prolonged conversations should be possible with only minimal listening fatigue.   This is the minimum requirement.

Listening tests have been made at various laboratories over the years, using many thousands of speakers.   It is interesting to note that slightly different frequency spectra are obtained when the speech pattern of the European male and female are analyzed, as shown in Fig. 2-2.   Children and speakers from non-European countries (e.g., Japan) exhibit different characteristics.

The internationally accepted telephonic specification has the following characteristics: a frequency response of 300 to 3400 hertz (Hz), harmonic distortion better than 26 decibels (dB), and a signal-to-noise ratio better than 30 dB.   These figures are by no means equivalent to those offered by high-quality audio equipment, which typically has a flat ±0.1 dB frequency response over the range 20 to 20,000 Hz. Once again, as in all areas of engineering, a compromise between cost, necessity, and the technologically feasible has been forced upon us.   Clearly, it would be desirable to have 300 million high-quality sound telephone links in operation, but is the additional cost reasonable?   Obviously not, and for this reason we must compromise.

Surprisingly, perhaps, telephones are also used for transmitting signals other than speech.   By use of a voice-frequency modulator-demodulator[1] (modem) and an acoustic coupler[2] it is possible to convey data messages via the telephone network.   The data pulses are converted into voice-frequency tones by the modem, in which form they are suitable for transmission.   The introduction of such equipments has imposed the need for additional constraints to be placed on the quality provided by a telephone channel.

---

[1] A unit that modulates a message signal enabling it to be transmitted, and also performs the reverse process.

[2] A unit that injects and/or receives acoustic signals from the telephone handpiece.

The human ear is insensitive to small changes in the phase of a signal; machines are not. Consequently, although small variations of phase will not affect the intelligibility of transmitted speech, such imperfections cannot be tolerated for the transmission of data. For this reason the maximum permitted variation in the group delay (phase) of the transmitted signal is carefully specified. Harmonic distortion may also affect data reception, yet remain imperceptible to the ear. Consequently this too must be carefully controlled.

Finally it is important to note that within a switched telephone network large ranges in speech volumes frequently occur. Not only must both weak and powerful speakers be catered for, but also differences in attenuation due to different transmission path lengths. We note, for example, that a telephone call may cross many international boundaries, or remain within the confines of a small local network. For these reasons the transmission system must accept a range of amplitudes greater than 30 dB.

## 2.3  TRANSMISSION REQUIREMENTS

The solution to a given transmission problem is dependent on several geographical factors. The terrain, the distance involved, and the number of channels required will simultaneously influence the chosen transmission method. We shall now see how today's telephone networks, which are based on FDM techniques, have accommodated these variables.

The transmission solution within a city, between a city and a town, and between two distant cities will all be different. In each case the distance of transmission and the number of telephone channels required will differ. Conventional telephony practice identifies the need for three types of link, and they are referred to as *local, junction,*[3] and *trunk*[4] connections, as shown in Fig. 2-3.

A *local* connection is one that joins a telephone subscriber to the local exchange. The distances involved are short.

A *junction* connection is one that joins a local exchange to a switching center. In Europe,[5] the most usual distance for this type of route ranges between 20 and 50 km, while the maximum distance is about 80 km.

A *trunk* connection is one that interconnects two switching centers, both nationally and internationally. These are the long-distance (maximum: 25,000 km), high-capacity routes.

Two types of telephone exchange have been identified within the simplified network illustrated in Fig. 2-3. The local exchange switches all calls within its local network (route 1), and passes all long-distance calls to the nearest switching center (route 2, 3). The switching center either passes the call to another local exchange within its area (route 2) or passes it to a distant switching center (route 3). Subdividing the switching tasks and keeping the number of direct transmission paths to a minimum greatly reduces the overall network costs. (Notice that no direct transmission path exists between points *A* and *B* in Fig. 2-3.)

We note that the complete transmission path between any two subscribers always includes two local connections, and may include one or more junction or trunk connections in addition. Clearly, careful attention must be given to the attenuation,

---

[3] This is the British term. In America the term *toll-connecting trunk* is used.

[4] This is the British term. In America the term *intertoll trunk* is used.

[5] In America the most usual distance for a toll-connecting trunk is of order 20 to 50 km also, while the maximum is about 300 km.

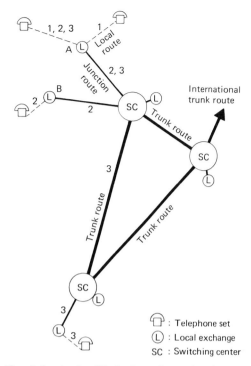

**Fig. 2-3**  A simplified view of a national telephone network.

or gain, introduced by these routes, since we cannot allow the speech volumes received at the subscriber's handpiece to be dependent on the transmission distance.

Since any transmission system contains gains, due to amplification, and losses, due to cable attenuation, a signal will have different levels at different points within the system, as shown in Fig. 2-4.  Conventionally, levels are expressed at different points with respect to a chosen position, known as the *zero reference point*.  The relative level of all other points within the system may then be denoted by the suffix

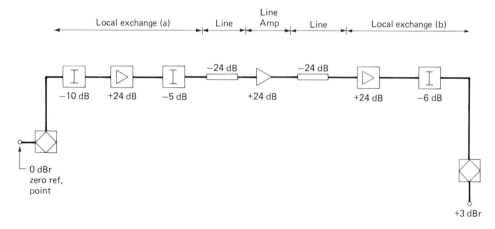

**Fig. 2-4**    Example of relative signal levels within a transmission network.

dB*r*, as shown in the diagram. This value is of course equal to the algebraic sum of the gains and losses between the point of interest and the reference point.

The signal seen at the reference point normally has a level of 1 milliwatt (mW), and the power of signals relative to this value is denoted by the symbol dB*m*. Thus 1 watt (W) is equivalent to 30 dB*m*.

Frequently it is convenient to express signal levels in terms of the corresponding level at the reference point; this is denoted by dB*m*0. Therefore,

$$\mathrm{dB}m0 = \mathrm{dB}m - \mathrm{dB}r \tag{2.1}$$

As an example, if a signal is found to have an absolute level of $-4$ dB*m* at a point where the relative level is $-9$ dB*r*, the signal at the reference point is $+5$ dB*m*0.

The junction and trunk routes are always arranged in such a way that they impose 0-dB loss or gain from end-to-end. Small gains may be corrected by introducing a passive network of known attenuation. The loss provided by such a network is known as the *insertion loss*. Alternatively, if in a similar way a corrective amplifier had been added, the gain provided is referred to as the *insertion gain*.

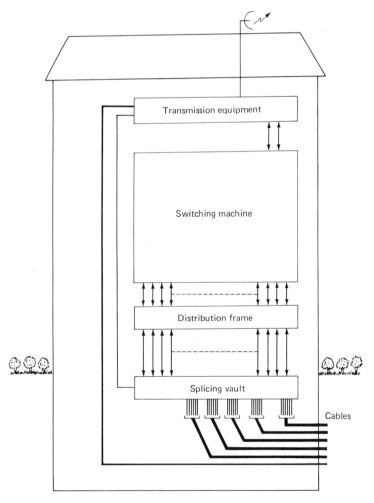

**Fig. 2-5** Layout of a local telephone exchange.

Only the local routes can significantly alter the overall attenuation, which is normally not allowed to exceed 5 dB per local connection.    Therefore we can see that it is the local network which exerts the dominant effect on the quality of a telephone call, not the long-distance transmission systems.    This is dictated by the economics of the situation.

The equipment required to provide long-distance communication, although expensive, is in continual usage, and may be optimized to advantage.    On the other hand, the local network consisting of many individual telephone sets, connection wires between telephone and exchange, and subscriber designated exchange plant is rarely used.    Any improvement must be made on a per subscriber basis and is expensive.

The layout of a local telephone exchange is shown in Fig. 2-5.    Each telephone in the local network is connected to the exchange via a single pair of wires, which is used for ringing the telephone bell, conveying the dialing pulses, and carrying the transmitted and received voice signals.    Separation of the two signals is accomplished by a resistive hybrid circuit which is included within the telephone set.

The single pairs of wires emerging from several telephones are grouped together and routed alongside each other within a single cable sheath.    The cables that are used must withstand a hostile environment, and have traditionally been provided with a protective lead cover.    Cables of this type enter the exchange within the *splicing vault,* where they are joined to smaller, more manageable plastic-covered cables.

Each spliced pair is connected to a *distribution frame,* whose purpose it is to allocate a dialing number for each telephone within the local network.    Physically, the distribution frame is an extremely large matrix which has connections to both the splicing vault and the switching machine.

The *switching machine,* under the direction of the dialing pulses, may either interconnect two cable pairs within the same exchange (local call) or connect one cable pair to the transmission equipment for a conversation requested with a distant exchange.

## 2.4  VOICE-FREQUENCY TRANSMISSION

A single pair of wires is usually the cheapest method of transmitting a voice-frequency message between two terminals, provided the distance involved is short (see Fig. 2-6a).    Such a system is limited by the cable capacitance of the interconnection pair, and a distance of approximately 8 km may be regarded as a maximum.

When a distance greater than 8 km is required,[6] loading coils must be inserted at regular intervals between the terminals (see Fig. 2-6b).    The *loading coils,* or inductors, compensate for the capacitance of the cable, which typically is in the range 30 to 50 nanofarads per kilometer (nF/km).    A curve showing the typical attenuation with frequency for a loaded, and unloaded cable pair is shown in Fig. 2-7.    The attenuation per kilometer[7] for an unloaded cable pair is approximately 0.5 dB at 800 Hz; when loaded the attenuation drops to approximately half this value.

Provided that the attenuation between the two terminals does not exceed 5 dB, a single loaded cable pair may be used to carry the voice-frequency signal.    When the attenuation exceeds this value, amplification is necessary in order to restore the

[6] The precise figure depends on the wire gage used and national practices.

[7] I consider here attenuation at voice frequencies.    The attenuation at 1 megahertz (MHz), (the half-bit rate frequency for 30-channel PCM) is typically between 20 and 32 dB.

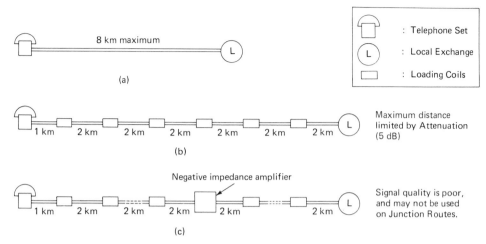

(a)

(b)

(c)

Maximum distance limited by Attenuation (5 dB)

Signal quality is poor, and may not be used on Junction Routes.

**Fig. 2-6**  Two-wire (single-pair) transmission systems.

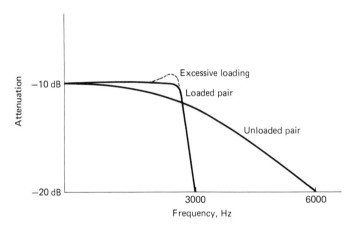

**Fig. 2-7**  The attenuation with frequency of loaded and unloaded cables.

signal to its original level.  Negative impedance bidirectional amplifiers are used, and are inserted at some of the loading coil sites (see Fig. 2-6c).  The signal quality provided by this system is poor, and consequently is not suitable for junction connections.  The degradation of signal is typically due to reflections within the cable produced by variations in the cable impedance.  It should be noted that since a negative impedance amplifier is used, a change in cable impedance with temperature will cause a change in the gain across the route.

An improvement in the transmission quality may be obtained by using four wires rather than two.  One amplified and loaded cable pair is used for the transmit direction, while a second is used in the receive direction (see Fig. 2-8).  The quality obtained with this configuration is suitable for junction connections.

The construction and parameters of voice-frequency cables are worthy of consideration.  The cable pairs consist of copper conductors that are separated from each other and other pairs by a paper insulator.  The overall construction is aimed at cheapness, and while suitable for voice-frequency communication, such cables are a poor medium for wideband signals (see Chap. 8).

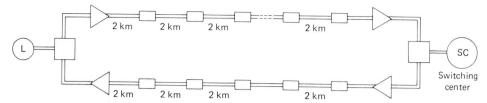

**Fig. 2-8**  Four-wire (double-pair) transmission system.

## 2.5  FREQUENCY DIVISION MULTIPLEXING

When many telephone channels are to be transmitted between two distant terminals, it is uneconomic to transmit each channel separately. Multiplexing techniques enable tens, hundreds, or even thousands of channels to be electronically combined in such a way as to provide one signal, which may be conveyed via a single transmission link. The most widely adopted multiplexing technique used in telephony today is frequency division multiplexing (FDM).

Suppose that $n$ telephone channels, each having a bandwidth $f_b$ of value 4 kHz are to be multiplexed by the technique of FDM. The multiplexer is arranged to modulate the telephone signals taken consecutively with carriers of frequencies $f_1$, $f_2, \ldots, f_n$. The modulated signals will contain both positive and negative sidebands, as shown in Fig. 2-9. Transmission of both sidebands would use up precious bandwidth, and consequently one of them must be removed by filtration. If the values of the carrier frequencies are chosen in such a way that each carrier is separated from the next by an amount at least as large as $f_b$, and the resultant modulated signals are mixed together, then a continuum stretching from $f_0$ to $f_n$ will be obtained (see Fig. 2-10). Each channel has been allocated a place in the frequency domain.

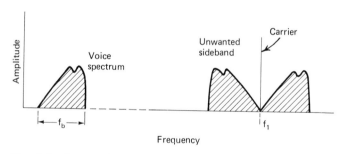

**Fig. 2-9**  Spectra illustrating modulation by a carrier.

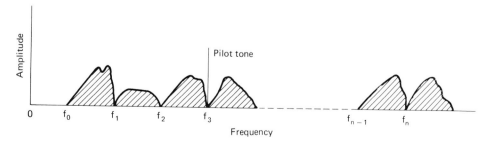

**Fig. 2-10**  The spectrum of a frequency division multiplexed signal.

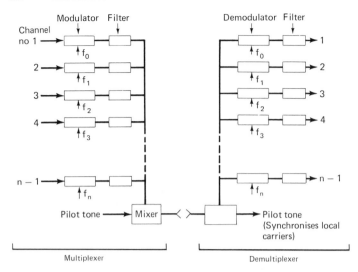

**Fig. 2-11**   The layout of a frequency division multiplex.

Balanced modulators are used, and consequently the carriers are automatically suppressed. However, at one point within the spectrum a special narrow bandwidth tone is inserted. The tone is referred to as the *pilot signal*[8] and is used at the demultiplexer for synchronizing the demodulator oscillators to the incoming signal. Thus any inaccuracies in the carrier frequency at the demultiplexer site are compensated for.

Separation of the individual channels from the composite signal is performed by using synchronous demodulators, one for each channel. These devices beat the composite signal, together with a signal equivalent to the original carrier, and produce a beat signal which after filtration is identical to the telephone channel input (see Fig. 2-11).

Thermally stabilized crystal-controlled synchronized oscillators with an accuracy of 1 part per million (ppm) are needed to generate the carrier signals. These devices are extremely expensive, but fortunately their cost may be shared between several multiplexers since the carrier frequencies remain identical. The filters used to remove the unwanted sidebands (multiplexer), and those used to filter the demodulated channel signals (demultiplexer) cannot be shared. They are difficult to design and are extremely expensive.

The design of an FDM equipment is based on analog components which have parameters that are liable to change with age. Consequently, in order that the quality of transmission may be kept within reasonable limits, periodic testing is required. In recent equipment designs, however, the service interval has been considerably extended.

The bandwidth requirement for an FDM equipment designed to combine 12 telephone channels stretches from 60 to 108 kHz and uses a transmission media which may be either a deloaded audio cable, open wire, or special pair cable. This multiplex provides the first order within the FDM hierarchy, and is the basic unit from which wideband signals representing larger numbers of channels are built up. The subsequent

---

[8] Conventionally, in radio communication practice the pilot is provided by attenuating the carrier signal. In FDM telephony the pilot occurs at a slightly different position to the carrier.

Table 2-2  Levels in the FDM Hierarchy

| Number of telephone channels | Bandwidth occupied, kHz | Name of group | Pilot tone, Hz |
|---|---|---|---|
| 12 | 60–108 | Primary group | Either 84,080 |
| 60 | 312–552 | Super group | or 84,140 |
| 300 | 812–2,044 | Master group | |
| 900 | 8,516–12,388 | Super-master group | |

levels in the FDM hierarchy after the *primary group* are referred to as *super, master,* and *super-master* groups; these are identified in Table 2-2.

The primary group systems are used extensively on existing audio pair cables. The loading coils at 2-km intervals are removed, and repeating amplifiers inserted every 8 km. It should be stressed that multiplexing equipment of this type is expensive and provides by far the major contribution to the total cost if only a short transmission distance is involved. The use of such systems must therefore be limited to cases where insertion of a new cable presents some difficulty, and to long-distance transmission. Currently, for transmission distances of less than 30 km it is more economical to use many four-wire circuits, or PCM.

## 2.6  CARRIER SYSTEMS USING COAXIAL CABLE

Coaxial cable carrier systems are commonly used when long-distance and high channel capacity are required, since it is no longer feasible to use paired cable if large bandwidths are involved. The spacing between the repeating amplifiers is a function of the cable attenuation and the system bandwidth such that the gain provided compensates for the loss introduced by the cable. Furthermore the gain/frequency characteristic must be arranged to be equivalent to the loss/frequency characteristic of the cable; this is known as *equalization* (see Chap. 8).

Unfortunately the attenuation of coaxial cables varies with temperature at the rate of approximately 0.2 percent per °C, depending on the actual cable used. In many countries the temperature variation between winter and summer can be as much as 20°C, which will cause a considerable change of cable loss. This must be allowed for within the design of the transmission system.

One way of overcoming the problem is to equip certain repeaters with automatic gain control circuits. This circuit measures the level of the pilot within the incoming signal and arranges the amplification provided by the repeater to be just sufficient to return the pilot level to some specified value. We shall see later that similar circuits are included within digital repeaters.

Modern repeater circuits use solid-state devices, and therefore have a low power consumption. This is typically in the range of 1 to 2 W per bidirectional repeater. For economic and practical reasons it is infeasible to provide a separate power supply for each repeater site; instead the required power is supplied via the coaxial cable. This technique is known as *power feeding*.

Most power feeding arrangements use a low current–high voltage combination. The maximum voltage that may be used is determined mainly by safety considerations for the maintenance staff, and lies in the range 250 to 1000 volts (V). It is interesting to note that the power feeding applied to submarine repeaters is somewhat similar.

However, because the distances are much longer, the applied voltage may be as high as 20,000 V!

If a fault occurs within a repeater or the cable, it is important that the location of the breakdown is found as soon as possible. Clearly it is impracticable to send maintenance crews to each repeater site along the transmission path. Consequently supervision and fault location circuits must be included within the repeater design. These circuits must be reliable and consume very small amounts of power. Their function can be roughly identified as:

1. Detect the occurrence of a fault within the repeater, or previous section of cable.
2. Indicate the exact location on the supervision control point, at the terminal.

Many different and ingenious supervision techniques have been devised. Some systems require a separate supervisory pair to be provided in addition to the coaxial cable, others do not. A few schemes permit the continuous monitoring of the carrier signal, without the need for taking the route out of service to check whether or not a fault has occurred. We shall return to this subject within Chap. 8.

## 2.7  CARRIER SYSTEMS USING RADIO

Radio transmission has an important place within any modern telecommunication network. In highly developed countries it may complement cable transmission systems. Here radio systems promote a high degree of reliability for the network taken as a whole, since there is no significant correlation between faults occurring within the two transmission media. In developing nations, or anywhere that has particularly difficult terrain, radio may provide the only practicable means of telecommunication.

There are three types of radio transmission systems in common usage within the telephone networks of today. These are:

1. Line of sight, which are connected in tandem to cover the required distance
2. Troposcatter, for distances up to about 400 km
3. Satellite, for very long distances

We shall consider within this section each of these systems taken in turn.

### 2.7.1  Line-of-Sight Radio Relay

Links of this type are used in Norway for crossing fjords, in Italy for traversing mountain ranges, in Britain for complementing cable systems, and within the Amazon jungle of Brazil. Such diverse applications confirm the importance of this type of transmission medium.

Most radio-relay systems employ frequency modulation, using a carrier in the range 1 to 12 gigahertz (GHz). (1 GHz = 1000 MHz.) Such systems secure a good signal-to-noise ratio for line-of-site transmission over distances typically in the range 40 to 50 km. However, under certain circumstances a much greater separation may be used. This would be true for example if one of the radio stations is located on a mountaintop.

A typical relay station consists of two parabolic antennas which are mounted as high above the ground as possible. One antenna is used for receiving the incoming signal, the other for onward transmission to the next repeater site. The arrangement normally compensates for an attenuation of about 140 dB under free space conditions;

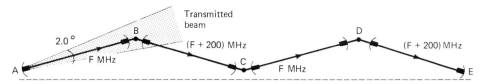

**Fig. 2-12**   Microwave transmission path.

however, this figure is subject to variations in the atmosphere by an amount dependent on the carrier frequency. For example, at frequencies greater than 10 GHz it becomes increasingly difficult to use microwave links through rain.

Since the receive and transmit antennas are mounted in close proximity to one another there is likely to be some coupling between them. If the coupling is sufficiently high, compared to the gain introduced between them, the arrangement could form a closed loop, and oscillate. For this reason the receiver and transmitter use different carrier frequencies, typically having a difference of about 200 MHz between each other. Consequently the carrier frequencies employed along a route are likely to be $f$, $f + 200$, and $f$ MHz for alternate hops, as shown in Fig. 2-12. For the reverse direction of transmission the complementary carrier frequencies are chosen in order to conserve bandwidth. Typically several transmitters operating at different frequencies share the same antenna. This is also true at the receiver.

The repeater sites are never positioned in a straight line along the transmission path, but as shown in Fig. 2-12. Otherwise the signal passed from $A$ and $B$ may during abnormal propagation conditions be received by $D$, and cause an interference with the signal from $C$. Since the parabolic antennas emit beams typically of width $2°$ at the half-power points, the displacement of the repeater sites from a straight line may be as small as $10°$.

Sometimes the received signal may be subject to fading. This is especially true for links that pass over water, where varying amounts of reflection can occur. In such cases a spare transmission path is used to provide a reserve capacity should the fading on the main route be particularly serious.

### 2.7.2 Troposcatter

Troposcatter systems enable transhorizon connections to be made by scattering the radio signals off the troposphere, which extends from the earth's surface to a height of about 10 km.

A narrow beam of signals is aimed at a point just above the horizon of the transmitter. The beam eventually reaches the edge of the troposphere, on the opposite side of the transmitter horizon, where it produces the equivalent of an illuminated spot due to scattering. The receiver uses a high-gain antenna to retrieve some of the scattered signals from the region of interest.

The scattering is caused by variations in the refractive index of the atmosphere, and is random. Consequently the signal-to-noise ratio varies considerably with systems of this type and the quality is normally lower than that offered by line-of-sight transmission. However, in many cases troposcatter[9] offers the only reasonable means of providing a transmission link.

[9] R. Gould and G. Helds, "Tropospheric Scatter Communications: Past, Present, and Future," *IEEE Spectrum*, vol. 9, 1972, p. 47.

### 2.7.3  Satellite

Commercial satellite communication began in the early 1960s with the Telstar project. However, since that date many more artificial satellites have been launched and international agreement on their usage reached. The frequency bands allocated by the International Radio Consultative Committee (CCIR) for satellite communication are:

$$\text{Earth to satellite:} \quad 5.925 \text{ to } 6.425 \text{ GHz}$$
$$\text{Satellite to earth:} \quad 3.700 \text{ to } 4.200 \text{ GHz}$$

In 1964 the International Telecommunications Satellite Consortium (INTELSAT), an international organization that provides and maintains satellite communication systems, was established. Since that date many tens of satellites have been brought into operation with hundreds of ground stations.

If an artificial satellite is placed in an equatorial orbit at a distance of approximately 36,000 km from the earth, to observers on the ground it appears stationary. Furthermore, every inhabited point on the earth may be reached by line-of-site transmission using only three geostationary satellite orbits.

The transmission requirement of a satellite is similar to the line-of-site microwave links described earlier; the power being supplied by solar cells mounted on the satellite structure. The earth stations must be capable of receiving very weak signals, and consequently characteristically have large extremely directional antennas associated with very low noise amplifiers.

At the present time the economic arguments in favor of satellite communications, compared to transcontinental cable systems, are confused. Currently, if many hundreds of telephone channels are to be transmitted over a transoceanic route, the costs of the two approaches are similar, although inherently the satellite approach should be cheaper. This is due to political factors rather than technical ones.

On the other hand, satellites are able to offer a cost advantage when the communication requirement is for a small number of channels between a large number of countries. To satisfy this need two types of satellite systems have been constructed; these are referred to as *multiple access* and *demand assignment,* respectively.

A frequency division multiple access (FDMA) satellite may for example serve eight different countries with 150 channels each. The eight earth stations are allocated different carrier frequencies, which are arranged such that they just fit the bandwidth of the satellite. The system enables each earth station to transmit the 150 channels to the eight different countries using its allocated transmit carrier, and in the receive direction retrieves the required 150 channels from the wideband signal (i.e., all eight carriers).

A time division multiple access (TDMA) satellite performs the above operation in the time domain. The eight earth stations use a common carrier and take turns in using this to transmit a sequence of pulses. Each sequence includes synchronization codes, station address codes, and the PCM equivalent of the 150-word-interleaved voice channels (see Chap. 6). Since the sequence length may be varied each frame depending on the particular traffic requirement at the time, TDMA equipments may be operated as demand assignment units.

A demand assignment satellite allocates channels to individual routes *only* when required to carry traffic. In distinction, a multiple access unit *always* preassigns channels to particular routes whether they are required or not. Thus, although multiple access satellites may be efficient for routes where the traffic is heavy, they are inefficient for routes where the traffic is light.

One problem that is common to all satellite communication systems is the time delay for a signal to pass from earth to the satellite, and return to earth again. This delay is approximately 270 milliseconds (ms). For television and other one-way transmissions there is no problem whatsoever; however, for two-way speech the delay (half a second) is just acceptable.

Sometimes extremely long transmission paths are required and two satellite links must be used in tandem. This occurs, for example, on connections between London and San Francisco, via New York. For such connections the CCITT recommends that one hop should be made by satellite, the other by cable. This avoids the very long and unacceptable time delays produced by using two satellite links in tandem.

## SUMMARY

We have described very briefly the main transmission methods that are used within many of today's telephone networks. Furthermore, we have seen how the switching function, the system economics, and geographical considerations dictate the best transmission solution.

The following books are particularly recommended for further reading:

1. Bell Telephone Laboratories Technical Staff: *Transmission Systems for Communications,* Bell Telephone Laboratories, Western Electric Co., 1970.
2. Flood, J. E. (ed.): "Telecommunication Networks," *Institution of Electrical Engineers Telecommunication Series 1,* Peregrinus, Stevenage, England, 1977.

# PULSE-CODE
# MODULATION

# 3

# Communication by Sampling

Earlier we introduced the concept that the identity of an amplitude modulated message signal may under certain circumstances be communicated by transmitting regular samples of the message, rather than the continuous signal. Here we shall define the term *sample* as the instantaneous measure of the amplitude value of a signal. The concept of communication by sampling is limited by the constraints imposed by the sampling theorem which we shall derive here.

There are many ways of communicating the measure of an amplitude, and there are as many pulse modulation schemes. We describe some of these in the first part of this chapter. In the later sections we aim to identify the limitations and constituent parts of a sampled communications system.

## 3.1 MODULATION SYSTEMS

Pulse amplitude modulation (PAM) is the classical example of an uncoded pulse modulation system. The message signal is uniformly sampled at specified time intervals. The sample values are transmitted by pulses, whose amplitudes are proportional to those of the message signal at the sampling instant. The relevant waveforms are illustrated in Fig. 3-1, together with a circuit representation.

PAM is frequently used as an intermediate step within time division multiplexers, as shown in Fig. 3-1. However, it is not usually considered as a signal for transmission over long distances.

Information about a message signal may be conveyed by varying the width of a pulse, pulse width modulation (PWM), or by varying its normal positions in time, pulse position modulation (PPM). The amplitude of the pulse is unrelated to the transmitted message in both of these cases.

A method for generating PWM and PPM samples directly is by comparison of the stored analog sample values with a sawtooth waveform, or ramp, as is shown in Fig. 3-2. The amplitude of the message signal is linearly proportional to the time interval ($t_{\omega 1,2}$) between the mean sampling instant and the instant when the message signal intersects the sloping[1] edge of the ramp. The time interval may be used to generate pulses that have a variable duration; this is PWM. Alternatively, a pulse that occurs at the instant of intersection may be generated; this is PPM.

---

[1] The slope of the reference sawtooth must be greater than the maximum slope of the message signal, otherwise more than one intersection can occur between samples (see Sec. 5.2.2).

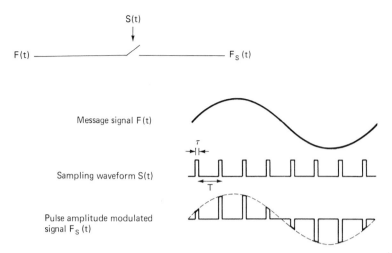

Fig. 3-1   The generation of a pulse amplitude modulated signal.

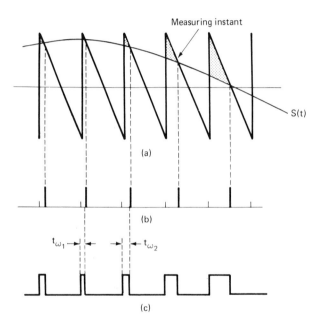

**Fig. 3-2**   The direct generation of PCM and PWM signals: (a) message signal $S(t)$, sampled by sawtooth waveform; (b) derived pulse position modulated (PPM) waveform; (c) derived pulse width modulated (PWM) waveform.

PWM and PPM are also not usually considered as a signal for transmission over long distances.   They may be used, however, as an intermediate step in the production of PCM signals.   In this case the PWM or PPM pulses may conveniently be employed to control a counting coder, as is described in Chap. 5 (Sec. 5.2.2).

PCM is the best-known, and today the most important pulse modulation system. Three separate operations are involved in providing a PCM signal, and each has been allocated a chapter in this book.   The first operation is to interrogate the message signal at regularly spaced intervals (sampling).   The second is to approximate the measured amplitude value to the nearest permitted voltage reference level, (quantizing; see Chap. 4).   The third operation is to represent the approximated (quantized) amplitude values as a series of pulses (coding; see Chap. 5).   This sequence of events is illustrated in Fig. 1-1, where the PCM equivalent of a message signal is also shown.

In PCM several pulses per sample are used to signify the amplitude value, instead of one pulse per sample as in the cases of PAM, PPM, or PWM.   Consequently PCM systems require an increased bandwidth for their transmission.   However, any small deformations in the height or width of the pulses are irrelevant since it is only necessary to know whether the pulse is present or absent in order to retrieve the original message.   Moreover, in a PCM transmission, noise is nonaccumulative because noisy PCM signals can easily be cleaned up, when it becomes necessary, by the process of regeneration.   Thus the quality of a PCM transmission is dependent on the sampling, quantizing, and coding processes, and not the length, nor the noise of the transmission media.   By contrast, the PAM, PWM, and PPM systems are continuously affected by noise and cannot be cleaned up, or regenerated.   The noise is accumulative, and the longer the distance of transmission the greater will be the noise.   This limits the maximum transmission distance for PAM, PWM, and PPM signals.

## 3.2  INFORMATION TRANSFER IN PULSE MODULATION SYSTEMS

It is instructive to compare the various pulse modulation systems in terms of information transfer.   We use the Hartley-Shannon law,[2] which states that if a transmission path is disturbed by white gaussian noise of power $N$, then the maximum information $I$ in bits per second that can be conveyed accurately by the transmission path is given by

$$I = B \, log_2 \left(1 + \frac{S}{N}\right) \tag{3.1}$$

where $B$ is the bandwidth of the channel, in Hz, and $S$ is the power of the transmitted signal.   In a real transmission there will always be some noise present; that is, $N$ will have a finite value.   Thus, the maximum information $I$ about a message signal that can be conveyed by a transmission path is always limited.

In PAM the message is transmitted using pulses that can assume any amplitude value, using in the ideal[3] case a bandwidth equivalent to the message signal.   However, although it is possible to generate an infinite number of amplitude values, the process of transmission will for a real system (that is, one that contains noise) make it impossible to recover the original infinite amount of information.   The maximum amount of information that can be transmitted will always be $I$ as defined by Eq. (3.1).   Thus information about the message signal will be lost during the process of transmission in a proportion dependent on the noise within the system.   For this reason PAM is not considered for transmission purposes.

---

[2] C. E. Shannon, "Communication in the presence of noise," *Proceedings of the Institute of Radio Engineers,* vol. 37, 1949, pp. 10–21.

[3] It is assumed here that the received message, retrieved from the PAM signal, is obtained using a rectangular characteristic low-pass filter.

In PCM each amplitude value is approximated to the nearest permitted level, which is represented by a set of pulses. Suppose that there are 16 possible (quantized) levels defined in a given system, and that the message signal of maximum frequency $f_m$ is sampled at $2f_m$. In this case the 16 possible levels may be represented by a sequence of 4 pulses, and the total information to be transmitted is $8f_m$. This information is finite, and can be better communicated via a transmission path at a comparatively low signal-to-noise ratio (see the discussion of error probability versus signal-to-noise ratio in Secs. 8.4.3 and 8.4.4).

With PCM the degradation that occurs in the complete transmission system is in the process of approximating the measured amplitude value. This is known as *quantization noise,* and can be reduced by increasing the number of permitted levels. It is interesting to note that if a sequence of 4 bits is available, then only 16 levels can be permitted; 5 bits provide 32 levels, 6 bits 64 levels, and so on. The more levels that are defined the less is the quantization noise, or approximation, and the greater is the bandwidth needed to transmit the extra pulses. I shall return to this subject in Chap. 4. Suffice it to state here that the signal-to-quantization-noise power ratio increases exponentially with bandwidth. This is an extremely efficient exchange of bandwidth for signal-to-noise ratio, and approaches the theoretical maximum attainable.

Tuller[4] has demonstrated that only coded systems, such as PCM, can efficiently exchange bandwidth for signal-to-noise ratio. The uncoded systems such as frequency modulation (FM) or PPM exhibit an improvement in signal-to-noise power ratio that is proportional to the bandwidth squared. This is much less efficient than the coded systems.

Let us return to our previous example, where there are 16 possible levels, and the message signal is sampled at $2f_m$ samples per second. I shall show later that in order to transmit $2f_m$ samples per second a transmission bandwidth equal to $f_m$ Hz is required.

If 4 pulses per sample are to be transmitted, as described above, then a transmission bandwidth of $4f_m$ Hz is required. For binary PCM, the signal-to-noise power ratio must be sufficient to distinguish between the presence or absence of a pulse; this requires $\sqrt{S+N}/\sqrt{N} = 2$.[5]

The same quantity of information ($8f_m$ bit/s) may be transmitted using a bandwidth equal to $2f_m$ Hz. This is achieved by transmitting 2 quaternary pulses per sample. A quaternary pulse can occupy any of four states, for example 1 V, 2 V, 3 V, and 4 V, compared to a binary pulse that has only two states. Thus 2 quaternary pulses may define an equal number of states as 4 binary pulses. However the signal-to-noise power ratio must be improved, and requires $\sqrt{S+N}/\sqrt{N} = 4$. This is an example of multilevel PCM.

These examples illustrate that for a given quantity of information it is possible to reduce the transmission bandwidth by increasing the average signal power and using a multilevel code. Clearly this is the reverse process of quantizing, and it is necessary therefore to increase the power exponentially for a linear decrease in the transmission bandwidth.

The choice between multilevel and binary PCM is ultimately dependent on the

[4] W. G. Tuller, "Theoretical Limits of the Rate of Transmission of Information," *Proceedings of the Institute of Radio Enigneers,* vol. 37, 1949, p. 468.

[5] J. M. Wozencraft and I. M. Jacobes, *Principles of Communication Engineering,* Wiley, New York, 1965.

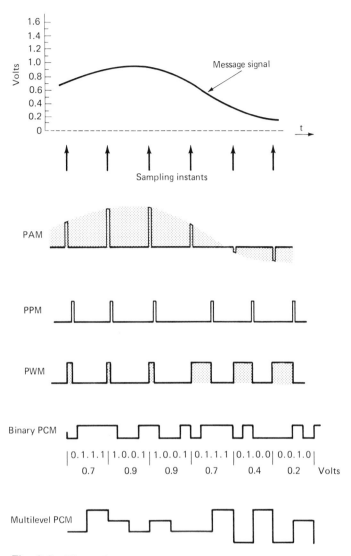

**Fig. 3-3** The various pulse modulation systems.

noise within the transmission system and on the practical difficulties of designing the coding/decoding equipments.

In conclusion to this section we refer the reader to Fig. 3-3, which illustrates the types of pulse modulation systems that have been described. In the remaining sections of this chapter we shall use **PAM** as the basis of discussion, since it is the easiest scheme to visualize. However, the points raised are equally applicable to the other pulse modulation systems.

## 3.3 SAMPLING SPECTRA

It is necessary, for a full explanation of the sampling process, to provide a theoretical basis. This we do here. The mathematical equations that describe the frequency

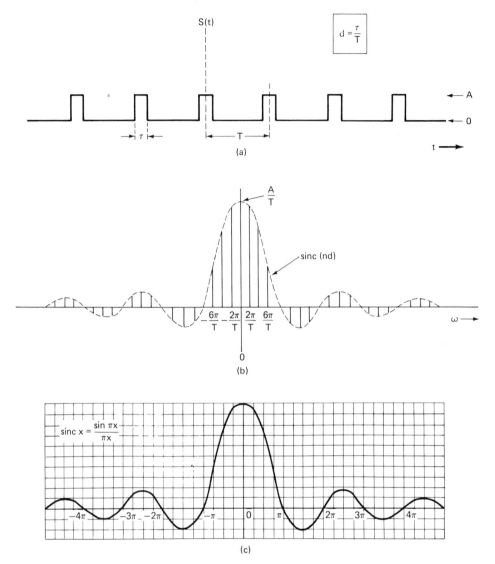

**Fig. 3-4** The spectrum of the sampling signal: (a) sampling waveform; (b) spectrum of sampling waveform: $S_n = Ad$ sinc $(nd)$; (c) the sinc function.

spectra of a message signal, a periodic series of pulses (the sampling waveform), and the PAM message signal will be derived.

It is apparent that the sampled signal $f_s(t)$ may be represented in terms of $f(t)$, the message signal, and $S(t)$, the sampling waveform, using the relationship:

$$f_s(t) = f(t)S(t) \tag{3.2}$$

The sampling waveform $S(t)$ has previously been considered as a series of impulses with an infinitesimal width, as shown in Fig. 3-1. This is referred to as *ideal,* or *instantaneous* sampling, and is physically impossible. It is more reasonable to assume that $S(t)$ is a periodic series of pulses of fixed amplitude, finite width $\tau$, and period $T$ seconds, as shown in Fig. 3-4a. This is known as *nonideal,* or *natural* sampling. The sampling waveform $S(t)$ is defined by

$$S(t) = \begin{cases} A & \text{when } -\tau/2 < t < \tau/2 \\ 0 & \text{when } \tau/2 < t < T - \tau/2 \end{cases}$$

The frequency spectrum of this waveform is not continuous: it exists only at certain discrete values of $\omega = 2\pi/T$. Thus, the spectrum of $S(t)$ may be graphically represented as a series of equally spaced lines, whose heights are proportional to the amplitudes of the discrete frequency components, as shown in Fig. 3-4$b$. It is instructive to take the exponential Fourier series of $S(t)$; using as the limits of integration $-\tau/2$ to $T - \tau/2$:

$$S_n = \frac{1}{T} \int_{-\tau/2}^{T-\tau/2} S(t) e^{-(jn2\pi/T)t} \, dt$$

$$= \frac{1}{T} \int_{-\tau/2}^{\tau/2} A e^{-(jn2\pi/T)t} \, dt$$

$$= -\frac{1}{T} \frac{AT}{jn2\pi} [e^{-(jn2\pi/T)t}]_{-\tau/2}^{\tau/2}$$

$$= \frac{A}{jn2\pi} 2j \frac{[e^{jn2\pi/T)\tau/2} - e^{-(jn2\pi/T)\tau/2}]}{2j}$$

$$= \frac{A}{n\pi} \sin\left(n\pi \frac{\tau}{T}\right) \tag{3.3}$$

Let the duty cycle $d = \tau/T$, and substitute into Eq. (3.3):

$$S_n = \frac{A}{n\pi} \sin n\pi d$$

This may be rewritten as

$$S_n = Ad\left(\frac{\sin n\pi d}{n\pi d}\right) \tag{3.4}$$

The spectrum represented by Eq. (3.4) can indeed only exist at certain discrete values, and has an envelope specified by the function in parentheses, as shown in Fig. 3-4$b$. This is the sinc function, as identified below, and in Fig. 3-4$c$. It occurs throughout the mathematics of sampling, and is well known to communications engineers and physicists[6] alike. If $x = nd$, then sinc $x$ is given by

$$\text{sinc}(x) = \frac{\sin \pi x}{\pi x} \tag{3.5}$$

The Fourier transform of a periodic function may be taken in the limit to be equivalent to the Fourier transforms of the individual components. It is possible to express a periodic function $s(t)$ with a period $T$ as

$$s(t) = \sum_{n=-\infty}^{\infty} S_n e^{(jn2\pi/T)t}$$

---

[6] The sinc function is sometimes referred to as the *top hat* function, since it is obtained from the Fourier series of a top hat, that is, a pulse. In physics the sinc function occurs for example, within optics in the mathematical analysis of Fraunhoffer lines. The function may be rewritten such that $\pi$ is omitted, sin $x/x$, this is called the *sampling function*, Sa($x$).

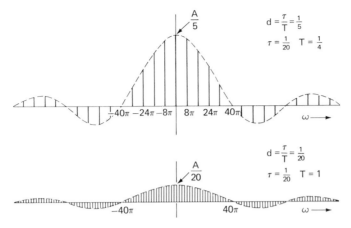

**Fig. 3-5**   The frequency spectrum of the sampling waveform $S(t)$.

But

$$\omega_0 = \frac{2\pi}{T}$$

Hence

$$s(t) = \sum_{n=-\infty}^{\infty} S_n e^{jn\omega_0 t}$$

Taking the Fourier transform on both sides gives:

$$\mathcal{F}s(t) = F \sum_{n=-\infty}^{\infty} S_n e^{jn\omega_0 t} \tag{3.6}$$

It can be shown that[7] the Fourier transform of $e^{jn\omega_0 t}$ is given by

$$\mathcal{F}(e^{jn\omega_0 t}) = 2\pi \, \delta(\omega - n\omega_0)$$

Substituting this result into Eq. (3.6):

$$\mathcal{F}s(t) = 2\pi \sum_{n=-\infty}^{\infty} S_n \, \delta(\omega - n\omega_0) \tag{3.7}$$

Substituting Eq. (3.4) into Eq. (3.7):

$$\mathcal{F}s(t) = 2\pi Ad \sum_{n=-\infty}^{\infty} \frac{\sin n\pi d}{n\pi d} \, \delta(\omega - n\omega_0) \tag{3.8}$$

This equation represents the frequency spectrum of the waveform $S(t)$, and consists of impulses that exist at $\omega = 0$, $\pm\omega_0$, $\pm 2\omega_0$, . . . , $\pm n\omega_0$.   The amplitude of the impulse located at $\omega = n\omega_0$ is given by $2\pi Ad$ sinc $(nd)$.   This spectrum is illustrated for $d = \frac{1}{5}$, and $d = \frac{1}{20}$, in Fig. 3-5.

We return to Eq. (3.2) in order to obtain an expression for the frequency spectrum of the sampled signal $f_s(t)$.   Using the convolution theorem it is possible to state that the multiplication of two functions in the time domain is equivalent to the convolution of their spectra in the frequency domain.   Thus Eq. (3.2) rewritten as the convolution of the two frequency spectra becomes:

[7] This is proved in B. P. Lathi, *Communication Systems*, Wiley, New York, 1968.

$$F_s(\omega) = \frac{1}{2\pi} F(\omega) * S(\omega) \tag{3.9}$$

Substituting Eq. (3.8) into Eq. (3.9) gives:

$$F_s(\omega) = \frac{1}{2\pi} 2\pi A d F(\omega) * \sum_{n=-\infty}^{\infty} \frac{\sin n\pi d}{n\pi d} \delta(\omega - n\omega_0)$$

$$= A d F(\omega) * \sum_{n=-\infty}^{\infty} \frac{\sin n\pi d}{n\pi d} \delta(\omega - n\omega_0)$$

$$= A d \sum_{n=-\infty}^{\infty} \frac{\sin n\pi d}{n\pi d} F(\omega) * \delta(\omega - n\omega_0)$$

$$= A d \sum_{n=-\infty}^{\infty} \frac{\sin n\pi d}{n\pi d} F(\omega - n\omega_0) \tag{3.10}$$

The graphic representation of Eq. (3.10) is illustrated in Fig. 3-6c. It is clear that the spectrum of the message signal $F(\omega)$ appears as a positive and a negative

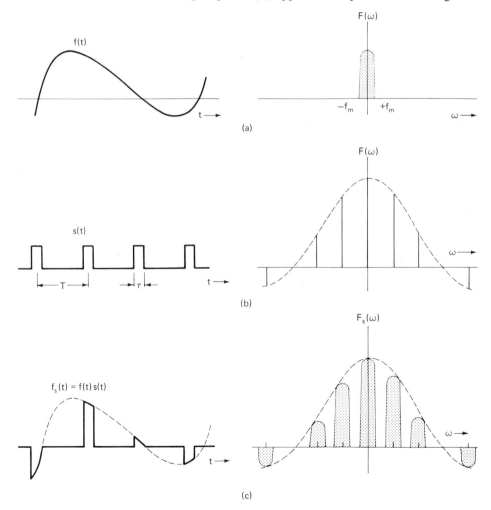

**Fig. 3-6** Waveforms and spectra for natural sampling: (a) message signal; (b) sampling signal; (c) pulse amplitude modulated signal.

sideband situated on either side of the impulses defined by Eq. (3.8). These impulses, it will be remembered, exist at 0, $\pm\omega_0$, $\pm2\omega_0$, $\pm3\omega_0$, . . . , $\pm n\omega_0$, and their amplitude varies as sinc $(nd)$. The *complete* information about the message signal is contained within each sideband, and is retrievable by filtering.

### 3.4  IDEAL SAMPLING

It is interesting to consider the limiting case of sampling when the sampling wave-form is a series of impulses that approach zero width ($\tau \to 0$, $d \to 0$). This special case is referred to as *ideal sampling*.

Since the value of sinc (0) is 1, then by substitution into Eq. (3.10), the spectrum for ideal sampling is given by

$$F_s(\omega)_{\text{Ideal}} = \frac{A}{T} \sum_{n=-\infty}^{\infty} F(\omega - n\omega_0) \tag{3.11}$$

The graphic representation of Eq. (3.11) is illustrated in Fig. 3-7, where the spectrum of the message signal $F(\omega)$ appears at the same locations as for natural sampling, while the amplitude remains constant.

The bandwidth required for transmitting an ideally sampled signal is infinite ($-\infty$ to $+\infty$). A naturally sampled signal, defined by Eq. (3.10), has a finite bandwidth and has, since $F_s(\omega)$ decays with frequency, a negligible energy level at high frequencies. It is clear that as the pulses are widened ($\tau$ increases), the spectrum decays faster, and the required transmission bandwidth becomes smaller. At first sight it appears that it is advantageous to maximize the pulse with $\tau$ in order to reduce the transmission bandwidth; however, it is necessary to consider the situation not only in the frequency domain, but also in the time domain.

If the pulse width $\tau$ is increased, then the time required to transmit the sampled signal is clearly also increased by the same amount. Normally, the sampled signals are used within a time division multiplex, and this requires $\tau$ to be small in order that as many channels as possible may be combined, or time interleaved between the sampling interval. We shall return to this problem later in this chapter, when I describe the principles of time division multiplexing.

### 3.5  SAMPLING THEOREM

We shall now show, using the mathematical results obtained earlier, that: "If a message signal $f(t)$ is band-limited to $f_m$ Hz then it may be completely characterized

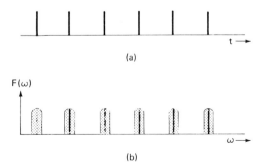

(a)

F($\omega$)

(b)

**Fig. 3-7**  Waveform and spectrum for ideal sampling: (a) sampling signal ($\tau = 0$); (b) sampled signal spectrum.

by samples taken at uniform intervals less than $\frac{1}{2}f_m$ seconds apart." This statement is known as the *uniform sampling theorem*.

The term *band-limited* means that there are absolutely no frequency components in its spectrum, above the frequency $f_m$ Hz. However, normal message signals do not have such a sharp frequency cutoff, and will contain frequency components at

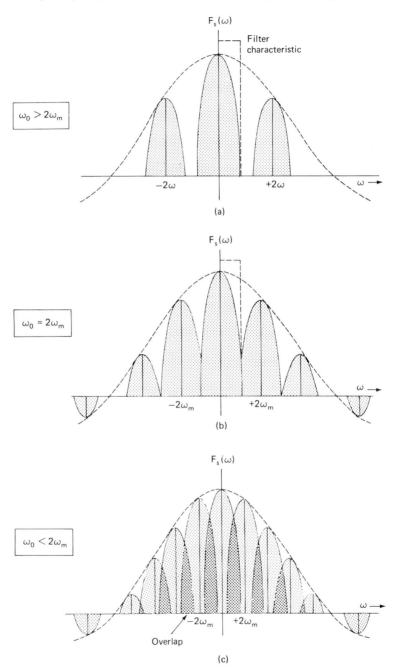

**Fig. 3-8** Sampling spectra for different sampling rates: (a) sampling at rate greater than Nyquist rate; (b) sampling at Nyquist rate; (c) sampling below Nyquist rate.

all frequencies. (This follows the Paley-Wiener criterion.) For this reason it has become normal practice to introduce sharp cutoff low-pass filters before sampling a message signal, to ensure that the band-limited condition is obeyed. We shall assume in the following description that the message signal is band-limited, and that the sampling pulses have a finite width (natural sampling).

The spectrum of the message signal $F(\omega)$ is repeated at intervals of $\omega_0$ radians per second within the envelope of the sampled signal $F_s(\omega)$, as is shown in Fig. 3-6c. The spectrum will repeat itself without overlap provided that $\omega_0 \geq 2\omega_m$, where $\omega_m = 2f_m$. This point is clarified in Fig. 3-8.

If $\omega_0$ is greater than $2\omega_m$ then the spectra are separated by a guard space, and appear as shown in Fig. 3-8a. In this case, it is possible to retrieve the original message signal, of bandwidth $f_m$, by filtering one positive sideband, as shown. Since the complete information about the original message is contained in *each* sideband, the negative sideband could equally well have been chosen.

If $\omega_0$ is equal to $2\omega_m$ then it is still possible to separate the message information by filtering one of the sidebands (Fig. 3-8b). This is the limiting case, since if $\omega_0$ is less than $2\omega_m$ (Fig. 3-8c) it is no longer possible to retrieve the original information by filtering. Thus, it is clear that the minimum sampling rate that may be used is $2f_m$. This is called the *critical sampling rate,* or *Nyquist rate* (after H. Nyquist of the Bell Telephone Laboratories).

It should be noted that the Nyquist rate is the absolute theoretical minimum. (This statement is true for uniform sampling *without* phase information.) Practical systems employ a much higher sampling rate in order to provide a large guard space (Fig. 3-8a), and therefore to simplify the filter design. As an example, the bandwidth occupied by a single telephone channel is 3.4 kHz; therefore the Nyquist rate is 6.8 kHz. However, practical systems such as the Bell T1 PCM system and the European PCM systems employ a sampling frequency of 8 kHz. Some switching machines, where transmission bandwidth is unimportant, sample at a frequency of 12.5 kHz, since this still further reduces the cost of filtration. For example, the electronic switching systems: ESS no. 101, produced by Western Electric, United States, and the SG1, produced by Northern Electric Company, Canada.

## 3.6  ALIASING ERROR (FOLD-OVER DISTORTION)

The importance of a truly band-limited signal can be seen by analyzing Fig. 3-8 in more detail. If the message signal contains components greater than $f_m$ it is clear that the spectral density function $F(\omega)$ will broaden to include these components, and the bandwidth is effectively increased to some new value $f'_m$. If $f'_m$ is so large that it causes a serious overlap of the spectrum (see Fig. 3-8c), then retrieval of the original message will become impossible. However, if $f'_m$ $(\omega f'_m)$ produces only a slight overlap (see Fig. 3-9), then the retrieved message signal will be slightly degraded. This is known as *fold-over distortion,* or *aliasing error.*

In the previous section, we described how the message signal may be separated by filtering one positive sideband and using an extremely sharp cutoff low-pass filter. If successive spectra overlap, then the filtered spectrum will be distorted due to two causes: (1) the loss of the tail beyond $\omega > \omega_m$, and (2) the addition of the inverted tail at the high-frequency end of the filtered spectrum. These distortions are illustrated in Fig. 3-9.

It is instructive to note that if a message signal is band-limited to 3.4 kHz and is

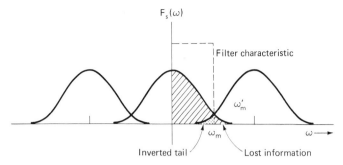

**Fig. 3-9** Aliasing error (fold-over distortion).

sampled as a rate of 6.6 kHz (Nyquist rate = 6.8 kHz), then the retrieved message signal will contain a strong component in the band 3.2 to 3.3 kHz and no components higher than this frequency.

## 3.7 INTERSYMBOL INTERFERENCE

Rectangular pulses will during their passage down a transmission path become dispersed, and spread out (see Sec. 8.2.3). [This is particularly true of long-haul fiber optic systems, where typically pulse distortion is a major problem (see Chap. 12).] Consequently, successive pulses will tend to overlap and cause a distortion known as intersymbol interference. This type of distortion occurs in PCM, or in time division multiplexed PAM, where the pulses of one channel interfere with adjacent channels. The phenomenon is best described by considering a series of binary pulses that represent a given amplitude value. The binary sequence 11011 is illustrated, as an example, in Fig. 3-10. In the diagram, the normally rectangular binary pulses have been modified by the filtration characteristics of the transmission system. This has produced pulses that have spread out, and caused an overlap into adjacent time slots.

Clearly, if the overlap is large then the decision as to whether or not a pulse exists at the sampling point may be incorrect. It should be noted that in practice there are likely to be contributions from several rather than just the nearest adjacent pulses as shown in the diagram.

There are two ways of reducing intersymbol interference. The first method, and the reason we introduce the topic here, is to separate the pulses by reducing the pulse width $\tau$ within the time period $T$. This is wasteful of bandwidth and causes, due to the Hartley-Shannon law, an impairment of the signal-to-noise ratio. The

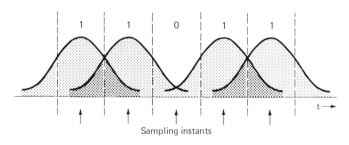

**Fig. 3-10** Intersymbol interference.

second method is to shape the transmitted signal, cable characteristics, and the receiver circuits in such a way that the interference is minimized. The signal waveshape is chosen to provide a maximum at the true sampling point, while it has a zero level at all other sampling points.

Normally the engineer is provided with a cable of given characteristics, and pulse shaping occurs due to this cable. Correction is provided by introducing filters (equalizers) that produce a final pulse shape that has minimal intersymbol interference. This subject is very important in line transmission, and will be described in more detail later (see Chap. 8).

## 3.8   A SAMPLING SYSTEM

A complete communication system that is based on sampling must have at least the elements shown in Fig. 3-11a. The PCM approach requires a few additional elements, which are identified in Fig. 3-11b, and are described in Chaps. 4 and 5.

The *input filter* ensures that the band-limited condition damanded by the sampling theorem is obeyed. The *sampling unit* interrogates the message signal at regular intervals T, using pulses of width $\tau$. The *output filter* separates the baseband message signal from its harmonics.

Earlier we made a comparison between ideal, and natural sampling. The result was obtained that, as one approaches the ideal case ($\tau \to 0$), then the amplitude of the harmonics of the message signal are virtually unattenuated at the high frequencies. Thus, the signal presented to the output filter will contain all the harmonics of the message signal at an equal amplitude. The purpose of the output filter is to remove these harmonics.

Intuitively, it is also clear that the output filter is required. Different waveforms, although of equal amplitude, can in fact be derived from the same samples, as shown

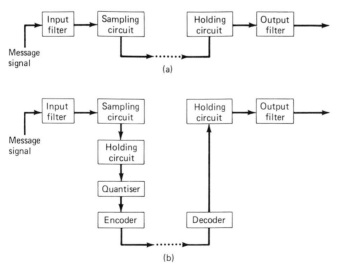

**Fig. 3-11**   The basic elements within a pulse modulation system: (a) minimum elements required in a sampling system; (b) minimum elements required in a PCM system.

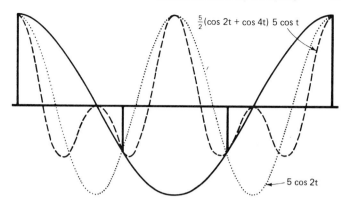

**Fig. 3-12**   The same samples, but different waveforms.

in Fig. 3-12. The output filter has a low-pass characteristic and retrieves the message signal without ambiguity by restricting it to a limited frequency range.

A sampling system based on PCM, or time division multiplexed PAM, requires an element known as the *sample and hold, or holding,* circuit. This circuit holds the amplitude value of each impulse for the duration of the sampling interval $T$.

The sample and hold circuit is used at the transmitter to give a PCM encoder sufficient time to perform a series of operations that result in the coded pulse pattern. At the receiver, the circuit is used to "stretch" the narrow impulses that are provided by demultiplexing the composite time division multiplexed signal (Fig. 3-13). The stretched signal is a rectangular wave which approximates the message waveshape (Fig. 3-13*d*).

The implementation of the sample and hold process is based on rapidly charging a capacitor to a potential which represents the sample amplitude. The narrow impulses that are derived from a low impedance source are allowed access to the capacitor by a switch, which is quickly open circuited toward the end of each impulse. Provided that the leakage from the capacitor is small, the sample values will be retained during the complete sampling interval $T$, whereupon the switch is closed once more, and a new impulse value is recorded.

The transfer function of the sample and hold circuit is that of a low-pass filter. This may be explained by considering a network with the transform $H(j\omega)$, as shown in Fig. 3-14. The input to this network is a pulse of width $\tau$, and amplitude $A$. The output pulse is also of amplitude $A$, but the width has been increased to a value $T$, where $T \gg \tau$.

The transform of an impulse with amplitude $A$ and width $\tau$:

$$\text{for } t = 0 \qquad\qquad\qquad = A\tau \qquad\qquad\qquad (3.12)$$

The transform of a pulse with amplitude $A$, width $T$:

$$\text{at } \left(-\frac{T}{2} < t < \frac{T}{2}\right) \qquad\qquad = AT\frac{\sin{(\omega T/2)}}{\omega T/2} \qquad\qquad (3.13)$$

The transform of the network $H(j\omega)$ is given by combining Eq. (3.12) and Eq. (3.13):

$$H(j\omega) = e^{-j\omega T/2}\frac{T}{\tau}\left[\frac{\sin{(\omega T/2)}}{\omega T/2}\right] \qquad\qquad (3.14)$$

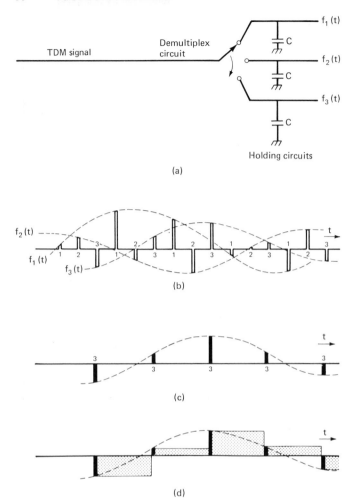

(a)

(b)

(c)

(d)

**Fig. 3-13**   The effect of the sample and hold circuit: (a) basic TDM switch; (b) composite TDM signal; (c) demultiplexed signal $f_3(t)$; (d) stretched signal after holding circuit $f_3(t)$.

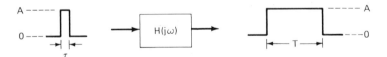

**Fig. 3-14**   The transfer function, associated with the sample and hold operation.

This of course represents the transfer characteristics of a perfect sample and hold circuit. The filtration is provided by the component in parentheses, which is the sinc function identified earlier in this chapter. Thus, at high frequencies because sinc $(\omega T/2\pi)$ tends toward zero, there is attenuation. The filter characteristic is identified in Fig. 3-15, where it is compared to the spectrum of the sampled signal. The two curves are similar in shape since they both contain sinc terms in their defining equations; however, the null points are different.

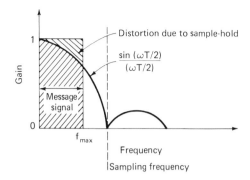

**Fig. 3-15**  The filtration characteristic of the sample, and hold function.

Careful analysis of Fig. 3-15 will reveal that there is some distortion in the range 0 to $f_{max}$, due to the unequal transmission of the spectral components in this range. Usually this distortion is small, and acceptable; however, if required, an equalizer that has the response $1/\mathrm{sinc}\ (\omega T/2\tau)$ may be added after the sample and hold circuit. The equalizer is here a passive circuit, which corrects for the distortion and yields an overall transfer characteristic that is flat.

It is also clear from Fig. 3-15 that when the sample and hold circuit is used at the receiver, the filtration is inadequate for retrieving the message signal information and suppressing the spectral harmonics. Some improvement in the filtration is obtained by increasing the sample rate, as shown in the diagram. However, an additional output filter is still required, and this ideally should have a tailored response such that the overall transfer characteristic is optimized. The desired characteristic is rectangular, as shown in Fig. 3-8$a$; however, this can only be approximated and in practice something less than ideal is acceptable.

We have assumed in the treatment of the sample and hold process that the sampling (switching) operation takes an infinitesimal time. Also, that the charging and discharging times of the capacitor are equally rapid. This is obviously not so. In fact, the optimization of a sample and hold circuit[8] is rather complex, and is beyond the scope of this book.

The implementation of the sampling operation has so far been considered as mechanical, while in practice electronic devices are used. A transmission circuit is required that has at its output during the sampling interval an exact reproduction of the input wave, and is otherwise zero. Such circuits are the subject of extensive literature,[9] and consequently, will not be duplicated here.

In this chapter we have identified several types of pulse modulation systems, and have shown PCM to be the most useful from the transmission point of view. In subsequent chapters the methods of implementing, and the theoretical limitations imposed by PCM transmission will be described.

---

[8] J. R. Gray and S. C. Kitsopoulous, "A Precision Sample and Hold Circuit with Subnanosecond Switching," *IEEE Transactions on Cable Television,* vol. CT11, 1964, p. 389.

[9] J. Millman and H. Taub, *Pulse, Digital and Switching Waveforms,* McGraw-Hill, New York, 1965.

# 4

# Analog to Digital
# Conversion—Quantization

This chapter is devoted exclusively to the subject of *quantization;* the name given to the process of approximating the individual message signal samples to the nearest permitted voltage reference level. The error introduced by this approximation, known jointly as *quantizing error* and *quantizing noise,* is usually the major impairment within a PCM transmission. Consequently the signal-to-quantizing-noise ratio will be considered in some detail throughout the chapter.

For the sake of clarity the chapter has been divided into distinct topic areas, each of which has in many cases several subdivisions.

The first section is a general introduction to the problem, and provides in addition the background mathematics needed later. We consider here the theoretical aspects of PCM transmission, and prove that coded systems may exchange signal-to-noise ratio with bandwidth most efficiently.

The second and third sections describe quantization schemes that take account of the characteristics of speech. Here special mention has been made of the systems in most common usage today.

The fourth and fifth sections examine the relationship between quantizing noise and the amplitude level of the message signal for different types of quantization scheme. The system impairments, other than quantization error, that may occur in practical equipments are discussed within the sixth section.

The concluding portions of this chapter are devoted to a special form of quantization known as *differential PCM,* or *delta modulation.*

## 4.1  UNIFORM QUANTIZATION

Suppose that a message signal can assume any amplitude value between 0 and $A$ volts. Also, that a reference scale is specified such that this scale extends from 0 to $A$ volts, in discrete uniform steps each of magnitude $\Delta v$ volts. Let the total number of steps be $S$. This is the situation illustrated in Fig. 4-1, where as an example, $A$ equals 1.6 V, $\Delta v$ equals 0.1 V, and $S$ equals 16.

The diagram shows that an impression as to the shape and size of the message signal may be conveyed by the approximated staircase-type waveshape. This signal either does not change, or it changes abruptly by an amount equivalent to one or more quantum steps, each of size $\Delta v$ volts.

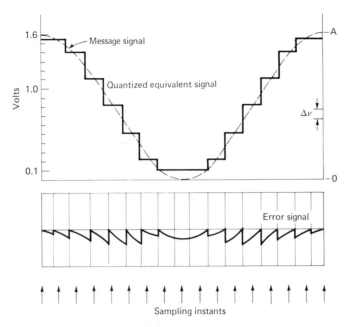

**Fig. 4-1**  Uniform quantization of a sine wave.

Since the step size $\Delta v$ is constant throughout the permitted amplitude range, the signal is said to be uniformly quantized.  We shall see later that more complicated systems exist that have a step size dependent on the amplitude level of the given sample (see Sec. 4.2).  However, throughout this section we shall consider $\Delta v$ to be of constant value.

### 4.1.1  Quantization Noise (Error)

If we return to Fig. 4-1 we see that there is a considerable error introduced between the original message signal and its quantized equivalent.

In practical terms, in the case of sound transmission this error will cause a continual background crackle.  If television transmission is considered, the error causes the number of gray tones that exist between black and white to be limited, while the picture appears generally noisy.

Clearly the approximation may be improved by reducing the step size $\Delta v$, and consequently increasing the number of steps $S$ that exist between the limits 0 and $A$ volts.  Eventually, provided $\Delta v$ is reduced sufficiently, the human ear or eye will be incapable of distinguishing between the original and the quantized signal.

In the diagram 16 quantum steps, or levels, have been defined.  In practical terms, such a small number of levels would provide a poor approximation of the message signal.  Subjective testing has established that studio quality color television can be conveyed using 512 levels, while 64 levels still produce a reasonable color picture.  Speech signals are of excellent quality if 128 levels are used, while intelligibility holds for as few as 8 to 16 levels.  In fact, 2-level quantization of speech is, with some difficulty, just understandable!

The importance of subjective testing has been emphasized several times already in this book.  During the design stage of an equipment, however, subjective analysis is not always possible, and for this reason it is usual to specify some easy to calculate and measure parameter as a design target. The tolerable value of this parameter is always determined by subjective experimentation.

A frequently used design parameter is the mean square quantizing error $\overline{E^2(t)}$. We shall use this parameter as a measure of degradation during the evaluation of different quantization systems later in this chapter. If $e$ represents the difference in amplitude between the actual signal and its quantized equivalent, then the mean square quantizing error may be expressed in terms of $\Delta v$ as shown below.

Since the step is $\Delta v$ volts, $e$ must exist between $-\Delta v/2$ and $\Delta v/2$. The quantum interval $\Delta v$ may itself be subdivided into $X$ small increments, each of width $\Delta v/X$ volts. We may assume that $e$ may occur with equal probability in the range $-\Delta v/2$ to $\Delta v/2$, and therefore at each subdivision $\Delta v/X$ (for speech signals, see Sec. 4.2).

Thus $e$ occurs with equal probability at: $0$, $\pm\Delta v/X$, $\pm 2\Delta v/X$, . . . , $\pm n\Delta v/X$, . . . , $\pm\Delta v/2$. Therefore,

$$\overline{e^2} = \int_{-\infty}^{\infty} e^2 p(e) \, de \tag{4.1}$$

where $p(e)$ is the probability density function of $e$, given by

$$p(e) = \begin{cases} \dfrac{1}{\Delta v} & \text{when } -\dfrac{\Delta v}{2} < e < \dfrac{\Delta v}{2} \\ 0 & \text{outside the above range} \end{cases}$$

By substitution in Eq. (4.1):

$$\overline{e^2} = \int_{-\Delta v/2}^{\Delta v/2} e^2 \frac{1}{\Delta v} \, de$$

$$= \frac{1}{\Delta v} \int_{-\Delta v/2}^{\Delta v/2} e^2 \, de$$

$$= \frac{(\Delta v)^2}{12} \tag{4.2}$$

However, provided the message signal is band-limited and sampled at a rate greater than or equal to the Nyquist rate, the mean square value of the error signal $E(t)$ is equivalent to the mean square value of its samples $e$.[1] Thus Eq. (4.2) may be rewritten as

$$\overline{E^2(t)} = \frac{(\Delta v)^2}{12} \tag{4.3}$$

However, conventionally we may consider the power of the quantization noise $(N_Q)$ equivalent to the mean square value of the error signal voltage difference. Thus the result is obtained:

$$N_Q = \frac{(\Delta v)^2}{12} \tag{4.4}$$

[1] For a rigorous analysis see: B. P. Lathi, *Communication Systems*, Wiley, New York, 1968, p. 135.

### 4.1.2  The Error Signal

We shall now consider the properties of the quantization error signal. According to the sampling theorem, the complete information about a message signal $f(t)$ may be conveyed by its samples, provided that the sampling rate is greater than or equal to the Nyquist rate $\omega_m$.[2] The sampled signal $f_s(t)$ is given by

$$f_s(t) = \sum_n f_n \delta(t - nT)$$

where $f_n$ is the $n$th sample of the message signal $f(t)$.

The message signal $f(t)$ is retrieved by filtering the signal $f_s(t)$, and suppressing all spectral components above the filter cutoff frequency $\omega_m$. The message signal may therefore be expressed as the time convolution of $f_s(t)$, and the filter characteristic. If we assume an ideal low-pass filter, which has a rectangular transfer function, then

$$f(t) = \sum_n f_n \delta(t - nT) * \frac{\sin(\omega_m t)}{\omega_m t}$$

By the convolution theorem:

$$f(t) = \sum_n f_n \frac{\sin[\omega_m(t - nT)]}{[\omega_m(t - nT)]}$$

$$= \sum_n f_n \frac{\sin(\omega_m t - n\pi)}{(\omega_m t - n\pi)}$$

$$= \sum_n f_n \, \mathrm{sinc}\left(\frac{\omega_m t}{\pi} - n\right) \tag{4.5}$$

However, the retrieved samples $f'_n$ are in error due to the process of quantization. The retrieved message signal $f'(t)$ is given by

$$f'(t) = \sum_n f'_n \, \mathrm{sinc}\left(\frac{\omega_m t}{\pi} - n\right) \tag{4.6}$$

Thus the error signal $e(t)$ is given by

$$e(t) = f(t) - f'(t)$$

$$= \sum_n (f_n - f'_n) \, \mathrm{sinc}\left(\frac{\omega_m t}{\pi} - n\right)$$

However, the instantaneous error $e_n = f_n - f'_n$. Therefore,

$$e(t) = \sum_n e_n \, \mathrm{sinc}\left(\frac{\omega_m t}{\pi} - n\right) \tag{4.7}$$

This result indicates that the spectral components of the quantizing noise exist only within the band limits of the message signal. Thus, it is justifiable to consider that all of the quantizing noise, defined by Eq. (4.4), is concentrated[3] within the baseband of the message signal. This is an important conclusion.

---

[2] This is a theoretical situation, not realizable in practice.

[3] The degree of concentration is largely dependent on the filtration characteristic.

Further examination of Eq. (4.7) reveals that, for periodic message signals, the fundamental frequency of the noise is equivalent to that of the message. The noise spectrum due to a periodic message signal occurs as a series of harmonics occurring within the band limit of the message signal.

However analysis of periodic signals is not very informative, since any result applies only to the specific case considered, nothing else. For this reason W. R. Bennett[4] assumed a band-limited random signal, in order to obtain a fairly typical result. Unfortunately the derivation of such results are somewhat involved and consequently they have been omitted from this book.

Finally, it is interesting to consider the effect of a less than ideal low-pass output filter on the output noise spectrum. Typically the transfer function of the output filter confines the baseband signal to a region something less than half the sampling rate. This will of course cause a reduction in the noise power. However, in practice this is not likely to exceed 1 dB.

### 4.1.3  Signal-to-Noise Ratio and Bandwidth Exchange

The ability of PCM systems to interchange signal-to-noise ratio with bandwidth was introduced in Chap. 3. We shall now investigate the mathematics of this exchange. It is first necessary to obtain an expression for the output signal power ($S_0$) of the system.

The output signal $f'(t)$ has already been defined by Eq. (4.6) as

$$f'(t) = \sum_n f'_n \, \text{sinc} \left( \frac{\omega_m t}{\pi} - n \right)$$

Conventionally the power of a signal is assumed to be equivalent to the signal amplitude squared. Now provided that the Nyquist criterion is obeyed, the mean square value of a band-limited signal is equal to the mean square value of the samples. Therefore,

$$S_0 = \overline{f'(t)^2} = \overline{(f'_n)^2} \tag{4.8}$$

The quantized samples $f'_n$ may assume any of $L$ finite levels, with an equal probability for each (for speech signals, see Sec. 4.2). Therefore the mean square value of $f'_n$ is given by

$$\overline{(f'_n)^2} = \frac{1}{L} [0^2 + (\Delta v)^2 + (2\,\Delta v)^2 + (3\,\Delta v)^2 + \cdots (L-1)\,\Delta v^2]$$

$$= \frac{(\Delta v)^2}{L} \sum_{n=0}^{L-1} n^2$$

$$= \frac{(\Delta v)^2}{L} \left[ \frac{L(L-1)(2L+1)}{6} \right]$$

$$= \frac{(\Delta v)^2 \, (L-1) \, (2L+1)}{6} \tag{4.9}$$

Since in practice many levels are specified $L \gg 1$, Eq. (4.9) reduces to

$$S_0 = \overline{(f'_n)^2} = \frac{L^2}{3} (\Delta v)^2 \tag{4.10}$$

---

[4] See W. R. Bennett, "Spectra of Quantized Signals," *Bell System Technical Journal,* vol. 27, 1948, p. 446.

By combining Eqs. (4.4) and (4.10) the output signal-to-noise power ratio is given as

$$\frac{S_0}{N_0} = 4L^2 \qquad (4.11)$$

It is now necessary to determine the signal-to-noise power ratio $S_i/N_i$ at the input to the decoder.

For binary transmission the decoder need only detect the presence or absence of a pulse. Since the presence ($A$ volts) or absence (0 volts) of a pulse occurs with equal probability, it is obvious that the average signal power $S_i$ is given by

$$S_i = \frac{A^2}{2} \qquad (4.12)$$

Clearly the value of $A$ must be chosen such that it is several orders larger than the average input noise level ($\Delta n$ volts). This reduces, but does not eliminate the probability of an error occurring within the detection process. We shall return to this subject in Chap. 8.

Let us assume

$$A = K \, \Delta n \qquad \text{where } K \text{ is a large number}$$

Now, the input noise power $N_i$ is given by $(\Delta n)^2$. Therefore, the input signal-to-noise power ratio becomes:

$$\frac{S_i}{N_i} = \frac{K \, \Delta n)^2}{2} \frac{1}{(\Delta n)^2} = \frac{K^2}{2} \qquad (4.13)$$

From Eqs. (4.11) and (4.13) the signal-to-noise-improvement power ratio is

$$\frac{S_0/N_0}{S_i/N_i} = \frac{8L^2}{K^2} \qquad (4.14)$$

This result must now be related to the transmission bandwidth requirement. Using binary transmission $\log_2 L$ pulses are required to represent $L$ quantum levels. The sampling theorem states that $2f_m$ samples per second are required to convey a message signal, which is band-limited to $f_m$ Hz.

Thus the transmission requirement $= 2f_m \log_2 L$ pulses per second. However, a transmission system of bandwidth $B$ can convey $2B$ pulses per second (see Chap. 8); therefore,

$$B = f_m \log_2 L \qquad (4.15)$$

Rewriting Eq. (4.15) in terms of $L^2$ gives:

$$L^2 = 2^{2B/f_m} \qquad (4.16)$$

Substituting Eq. (4.16) into (4.14) gives:

$$\left(\frac{S_0/N_0}{S_i/N_i}\right)_{\text{Binary}} = \frac{8}{K^2} 2^{2B/f_m} \qquad (4.17)$$

This result is significant because it shows that a small increase of transmission bandwidth results in a large increase in the signal-to-noise-improvement power ratio. This is true only of coded transmission systems such as PCM.

The expression defined by Eq. (4.17) may be generalized to take into account multilevel PCM transmission.  In this case, instead of transmitting the signal using only two logic states (pulse absent or present), the number of states is increased and hence the transmission bandwidth may be reduced.

If the number of logic state equals $M$, it can be shown that[5] the signal-to-noise-improvement power ratio is given by

$$\left(\frac{S_0/N_0}{S_i/N_i}\right)_{\text{Multilevel}} = \frac{24}{K^2(M-1)(2M-1)} M^{2B/f_m} \tag{4.18}$$

The results of Eqs. (4.17) and (4.18) have been obtained by considering positive going pulses only.  That is, in the binary case, only the states 0 volts and $A$ volts were allowed.

If such a signal were sent, it would be extremely wasteful of power since a direct-current (dc) component of value $A/2$ volts will also be transmitted.  The dc component serves no useful purpose and may be removed by sending the signal as $-A/2$ (state 0), and $+A/2$ (state 1).  In this case the input signal power $S_i$ may be reduced by an amount $(A/2)^2$.

The improved signal-to-noise power ratio, with dc components removed, is given by

$$\left(\frac{S_0/N_0}{S_i/N_i}\right)_{\text{Binary}} = \frac{16}{K^2} 2^{2B/f_m} \tag{4.19}$$

$$\left(\frac{S_0/N_0}{S_i/N_i}\right)_{\text{Multilevel}} = \frac{48}{K^2(M^2-1)} M^{2B/f_m} \tag{4.20}$$

It should be noted that if the input signal power is reduced by removing the dc component, the input signal-to-noise power ratio is also modified.  This becomes in the case of multilevel PCM:

$$\frac{S_i}{N_i} = \frac{K^2(M^2-1)}{12} \tag{4.21}$$

### 4.1.4   Theoretical Efficiency of a PCM System

Before considering practical systems, it is appropriate to consider the theoretical maximum efficiency of a PCM system by use of the Hartley-Shannon law, which states that

$$I = B \log_2\left(1 + \frac{S_i}{N_i}\right) \tag{4.22}$$

where $I$ = theoretical maximum information that can be transmitted
$B$ = transmission bandwidth
$S_i$ = input signal power
$N_i$ = input noise power

We saw earlier, during the derivation of Eq. (4.15), that $2f_m \log_2 L$ pulses per second must be transmitted in order to convey a message signal of bandwidth $f_m$, and $L$ quantum steps.  Therefore, the information to be transmitted $I'$ is

$$I' = 2f_m \log_2 L \tag{4.23}$$

---

[5] The derivation is similar to that for the binary case; see Lathi, *Communication Systems.*

If this amount of information is transmitted using multilevel PCM of $M$ levels, then Eq. (4.15) may be rewritten in the more general form:

$$B = f_m \log_M L \tag{4.24}$$

From Eqs. (4.23) and (4.24):

$$\begin{aligned} I' &= (f_m \log_M L)(2 \log_2 M) \\ I' &= B \log_2 M^2 \end{aligned} \tag{4.25}$$

We shall now derive an expression for the input signal-to-noise power ratio, which has already been defined in terms of $M$, by Eq. (4.21) rewritten below:

$$\frac{S_i}{N_i} = \frac{K^2(M^2 - 1)}{12}$$

Therefore,

$$M^2 = 1 + \frac{12}{K^2} \frac{S_i}{N_i} \tag{4.26}$$

By substituting Eq. (4.26) into Eq. (4.25):

$$I' = B \log_2 \left( 1 + \frac{12}{K^2} \frac{S_i}{N_i} \right) \tag{4.27}$$

$I'$ represents the information that must be conveyed in order to retrieve the quantized message exactly. In fact, if the signal-to-noise power ratio of the transmission path is $S_i/N_i$, and the bandwidth $B$, the maximum information that can be conveyed is $I$, given by Eq. (4.22). Therefore, for PCM transmission $K^2/12$ times as much signal power is required compared to that of an ideal system.

A comparison between the performance of an ideal system and various multilevel PCM systems is illustrated in Fig. 4-2, where $S$ is the number of levels.

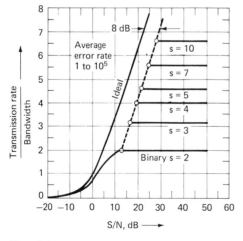

**Fig. 4-2** A comparison between ideal, and multilevel PCM systems. (From Bhagwandus P. Lathi, *Communication Systems,* Wiley, New York, 1968.)

## 4.2 NONUNIFORM QUANTIZATION

In the previous sections we have considered the case of quantization using a fixed step size $\Delta v$. Moreover in Eq. (4.4) we obtained the result that the quantization noise power is equivalent to $(\Delta v)^2/12$, provided that there is an equal probability of occurrence for any amplitude value within the permitted range. We shall now investigate the possibility of reducing the magnitude of this noise.

If we examine the cause of error as shown in Fig. 4-1, two methods of reducing the noise power suggest themselves. Clearly the noise will be reduced by simply making the step size $\Delta v$ smaller. Alternatively, a more exact description of the message signal may be obtained by increasing the sampling rate[6] to something greater than the Nyquist rate, which is assumed here. Unfortunately both of these solutions require a considerable increase in transmission bandwidth for only a small reduction of the noise power.

A third method of reducing the noise takes into account the statistics of the message signal. The probability distributions of message signal amplitudes are rarely if ever uniform; an assumption made in the derivation of Eq. (4.4). In speech, which we shall consider here, the probability of occurrence of a small amplitude is much greater than a large one. In fact, experiments have shown[7] that the probability distribution of speech signal amplitudes is approximately exponential. Consequently, it seems appropriate to provide many quantum levels ($\Delta v$ = low value) in the small amplitude region, and only a few ($\Delta v$ = high value) in the region of large amplitudes. In this case, provided the total number of specified levels remains unaltered, no increase in transmission bandwidth will be required. This technique is referred to as *nonuniform quantization*.

It is interesting to note that information theory tends to indicate that a system based on nonuniform quantization will provide a more faithful replica of the message. A coded signal carries the maximum information when each level has an equal probability of occurrence. This must be true since, in the limit, if a particular level never or always occurs then the information supplied by this level is zero. Consequently, a system designed on the basis of maximum information transfer will specify many levels in the small amplitude region, and only a few in the high.

At this juncture it should be pointed out that relating the quantum level positions to the probability distributions of signal amplitudes must be undertaken with caution. In the case of speech, the long-term amplitude probability distribution is exponential. However, individual speech sounds depart widely from the long-term mean, and due allowance must be made lest the nuances, if not the intelligibility of the spoken word, become degraded.

The range of amplitudes that can occur within a transmission system is considerable. Both loud and soft speakers over varying distances must be taken into account. Telephony, in particular, demands that quantizing equipment be designed to accept a wide range of amplitudes due to the enormous attenuation differences of transmission lines within a switched telephone network. (The difference in level between two telephonic-speech signals may easily exceed 30 dB.) It is clear that a nonuniformly quantized system is inherently better able to accommodate a larger range of signal

---

[6] For a quantitative treatment see W. R. Bennett, "The Spectra of Quantized Signals," *Bell System Technical Journal*, vol. 27, 1948, p. 446.

[7] W. B. Davenport, Jr., "An Experimental Study of Speech-Wave Probability Distributions," *Journal of the Acoustical Society of America*, vol. 24, 1952, pp. 390–399.

amplitudes than the uniform case, assuming of course that the total number of levels remains fixed.

Consequently, the usual design criterion of a nonuniformly quantized system is that the signal-to-quantizing-noise ratio should always be greater than some acceptable figure (e.g., 35 dB), over the greatest possible range of signal amplitudes (e.g., 30 dB).

### 4.2.1  Implementation of Nonuniform Quantization

Equipments employing a variable step size may be implemented in several ways, as shown in Fig. 4-3.

The equipment may be arranged to specify quantum levels of variable spacing directly. However, this approach is somewhat troublesome, due to the practical difficulty of maintaining accurately defined drift-free switching thresholds using standard analog components. Nevertheless, this problem has been overcome in several successful designs.

Alternatively the message signal may be uniformly quantized using an extremely small fixed step size over the entire range of possible amplitudes. The resultant quantum values may then be digitally coded and thereafter translated, or processed, to give a code which represents the nonuniform step positions. This technique will be described in more detail within Chap. 5.

In the third method of implementation the message signal is passed through a nonlinear network, called a *compressor*, which has an input-to-output transfer function as shown in Fig. 4-4. The distribution of signal amplitudes becomes modified, due to the attenuation of the large amplitude signals. Finally the compressed signal is uniformly quantized to produce the nonuniformly quantized signal. The operation of compression followed by uniform quantization is known as *companding*. A circuit that has a transfer function, which is the inverse of the compressor, is provided at

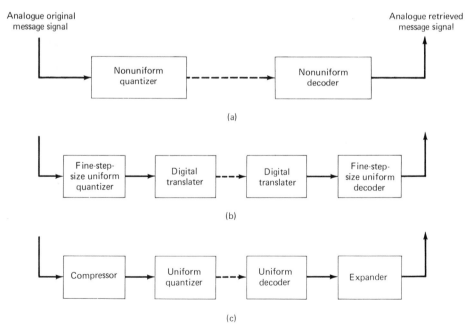

**Fig. 4-3**  The implementation of nonuniform quantization.

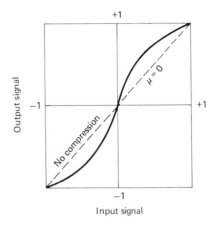

**Fig. 4-4** The compression produced by the $\mu$ companding law.

the receiver in order that the proper signal amplitude distribution is restored. This complementary device is referred to as an *expandor*.

Any of these three implementation methods may be used to provide an identical result. Most current systems tend to favor the usage of compander circuits. However, the availability of cheap custom-made integrated circuits may in future promote either of the first two approaches.

### 4.2.2 Companding Laws

The variation in step size with signal amplitude, that is the degree of nonuniformity, is defined by the chosen companding law. This law dictates the shape of the curve shown in Fig. 4-4, and is chosen to fit the statistics of the message signal within the limits already described.

Clearly both positive and negative signals must be handled with equality. Consequently, the companding curve must be skew symmetrical, and pass through the origin, as shown in Fig. 4-4. Furthermore, the portion of the companding curve which relates to the small amplitudes, that is the central portion, will have a steeper slope compared to that obtained by uniform quantization. The ratio of the two slopes is known as the *companding advantage*.

The best known of the companding laws is perhaps the $\mu$ law, described by B. Smith[8] in 1957. This is logarithmic in form, and is defined by the relationship:

$$V_c = \frac{a \log (1 + \mu V_i/a)}{\log (1 + \mu)} \qquad \text{for } 0 \leq V_i \leq a \qquad (4.28a)$$

$$V_c = \frac{-a \log (1 - \mu V_i/a)}{\log (1 + \mu)} \qquad \text{for } -a \leq V_i \leq 0 \qquad (4.28b)$$

where $V_i$ = input signal amplitude, in volts
$V_c$ = compressed signal amplitude, in volts

The operating region is defined by the value of $a$, which is given by

[8] B. Smith, "Instantaneous Companding of Quantized Signals," *Bell System Technical Journal,* vol. 36, 1957, p. 653.

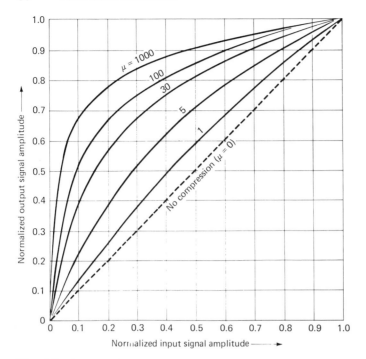

**Fig. 4-5**  Compression using the $\mu$ companding law, for different $\mu$ values. (From B. Smith, "Instantaneous Companding of Quantized Signals," *Bell System Technical Journal,* May 1957. Copyright 1957, American Telephone and Telegraph, reprinted by permission.)

$$(V_c)_{V_i=a} = a$$

The degree of compression may be varied by changing the value of the *compression parameter* $\mu$, as illustrated in Fig. 4-5. Here only positive excursions of the signal are considered, since negative values are handled in a similar manner. Furthermore in the diagram the variables $V_c$ and $V_i$ have been normalized, such that they each occupy the range 0 to +1. The convention of considering positive normalized excursions only, is standard practice and is used throughout this book.

Thus, the defining equation of the $\mu$ law, Eq. (4.28), may be rewritten as

$$V_c = \frac{\log (1 + \mu V_i)}{\log (1 + \mu)} \tag{4.29}$$

The shape of the curve traced by Eq. (4.29) is essentially logarithmic for large values of $V_i$. However, for small input signals, the expression reverts to an almost linear form, such that $V_c$ and $V_i$ become directly proportional to one another. The transition to linearity occurs where $V_i$ is somewhat greater than $\mu^{-1}$. Practical values of $\mu$ are usually chosen to be in the order of 100.

Any practical companding law must revert to a linear, or at least near-linear relationship for small input signals. Such signals are not only the most common, but also the most important for the intelligibility[9] of speech at constant volume. The natural-

[9] J. C. R. Licklider and I. Pollack, "Effects of Differentiation, Integration and Infinite Peak Clipping upon Intelligibility of Speech," *Journal of the Acoustical Society of America,* vol. 20, 1948, pp. 42–51.

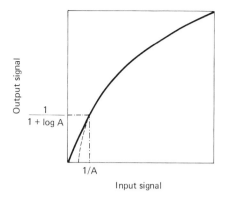

**Fig. 4-6** The compression produced by the A companding law.

ness of the spoken word may be associated with the large amplitudes, and the chosen companding curve must therefore take these into account.

Logarithmic companding laws such as the $\mu$ law tend to fit the above criterion very well. Nevertheless, one cannot avoid speculating about the possibility of obtaining a substantial improvement in companding advantage by adopting a nonlogarithmic characteristic. Several schemes have been proposed, notably in Japan where a law based on a hyperbolic-sine term has been suggested.[10]

However, although equipments based on these nonlogarithmic laws are slightly less complex, they generally accept only a narrow range of signal volumes. Moreover, theoretical analysis[11] will confirm that if the quantizing noise power is to be reduced to a level beyond that attainable with logarithmic companding, a smaller range of signal volumes, and consequent diminished naturalness, must be tolerated. For these reasons practical equipments have generally adopted logarithmic companding laws.

We have noted that the required compression is ideally linear at small amplitudes, but logarithmic at the large amplitudes. The $A$ law, proposed by K. W. Cattermole[12] in 1962, satisfies this criterion exactly, as shown in Fig. 4-6.

The normalized defining equations of the $A$ law are

$$V_c = \frac{A V_i}{1 + \log A} \qquad 0 \le V_i \le \frac{1}{A} \tag{4.30a}$$

$$V_c = \frac{1 + \log (A V_i)}{1 + \log A} \qquad \frac{1}{A} \le V_i \le 1 \tag{4.30b}$$

There are clear similarities between the $A$ and $\mu$ laws. The transition to linearity occurs where $V_i$ equals $A^{-1}$, while practical values of the compression parameter $A$ are usually in the order of 100. The main difference between the two laws is that the $A$ law is perfectly linear for small signal values, while the $\mu$ law only approximates linearity and no two quantum steps are exactly equal in size. However, practical

[10] H. Kaneko and T. Sekimoto, "Logarithmic PCM Encoding without Diode Compander," *IEEE International Convention Record,* part 8, 1963, pp. 266–281.

[11] Smith, "Instantaneous Companding."

[12] The first description of the $A$ law was given in R. F. Purton, "A Survey of Telephone Speech Signals Statistics, and Their Significance in the Choice of a PCM Companding Law," *Proceedings of the IEEE,* vol. 109B, 1962, p. 60.

equipments[13] using the $\mu$ law, and diode companders, have tended towards a superior linearity for small signal values than predicted by theory. This has been attributed to the perfectly linear diode response at the low levels.

Equipment employing the $\mu$ law is in use within the telephone networks of America and Japan. These countries use the Bell T1 system, which originally included the D1 primary multiplex[14] ($\mu = 100$) and currently the D2 primary multiplex ($\mu = 255$). The D2 multiplex utilizes a segmented companding law (see Sec. 4.3).

The telephone networks of Europe, Africa, South America, and all international routes utilize within the CCITT primary multiplex, the segmental form of the $A$ law, with the compression parameter $A$ equal to 87.6.

## 4.3   SEGMENTED QUANTIZATION (COMPANDING)

In the previous section we identified companding laws that have a smooth, regularly shaped characteristic. Such laws require an infinite number of quantum step sizes to be specified if the defined characteristic is to be met exactly.

The cost of implementing a nonuniform quantization system may be reduced considerably by limiting the number of quantum step sizes specified. The purpose of segmentation, then, is to combine the advantages of compression with reasonable cost, by approximating the smooth companding curves as a series of chords representing regions of uniform quantization, or fixed step size. The area defined by each chord is known as a *segment*, and hence the term *segmented companding*.

The $A$ law, defined by Eq. (4.30), is ideally suited to segmentation, and it is in this form that it is most used. The segmented $A$ law, as defined by the CCITT,[15] is used here as an example and is shown, for positive values only, in Fig. 4-7.

The CCITT equipments specify seven different quantum step sizes, equivalent to seven segments for both positive and negative excursions. In total, therefore, 14 segments are required. However since the slope of the central region is the same for both positive and negative values it is normal practice to consider the two segments about the origin as one. Consequently the requirement is for a total of 13 segments.

Analysis of Fig. 4-7 will reveal that in order to specify the location of a sample value it is necessary to know the following:

1. The sign of the sample (positive or negative excursion)
2. The segment number
3. The quantum level within the segment

We shall see within Chap. 5 that this information may be easily compiled and coded; however, here these details are omitted.

Intuitively it is clear that the approximation between the theoretical, smooth characteristic and its segmental equivalent should be almost constant throughout the permitted range of amplitude values. The linear central segment of the $A$ law suffers no approximation, provided that the slope of this segment is arranged to coincide with the theoretical curve. During the logarithmic portion of the characteristic, the approximation may be kept almost constant by making the ratio of the slopes between

---

[13] R. H. Shennum and J. R. Gray, "Performance Limitations of a Practical PCM Terminal," *Bell System Technical Journal*, vol. 41, 1962, pp. 143–171.

[14] International Telegraph and Telephone Consultative Committee, *CCITT Orange Book*, vol. III-2, G711 (recommendation), International Telecommunications Union, Geneva, Switzerland, 1977.

[15] Ibid.

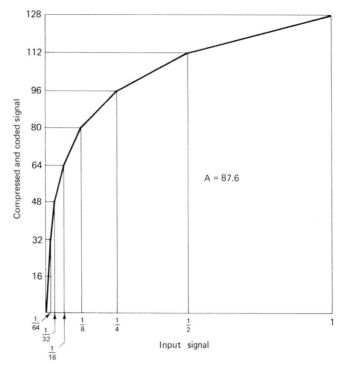

**Fig. 4-7** Segmented companding curve based on the A law.

adjoining segments equal. In this case each segment occupies the same logarithmic interval. It should be pointed out that each segment need not, but nevertheless usually does, contain an equal number of quantum steps.

The gradient of the slope within the first segment may be obtained by simply differentiating Eq. (4.30a), the defining equation of the A law, for small signal values.

$$\tan \alpha_1 = \left[ \frac{d(V_c)}{d(V_i)} \right]_1 = \frac{A}{1 + \log A} \qquad \text{for } 0 \le V_i \le \frac{1}{A} \qquad (4.31)$$

We note, as shown in Fig. 4-6, that if a total of 129 quantum steps are specified for positive excursions of the signal, and that if the first segment stretches over 32 of these steps, then $\tan \alpha_1$ is given by

$$\tan \alpha_1 = \frac{32/128}{1/64} = 16 \qquad (4.32)$$

Therefore, from Eqs. (4.31) and (4.32):

$$\tan \alpha_1 = \frac{A}{1 + \log A} = 16 \qquad (4.33)$$

The solution to this transcendental equation gives the result that the compression parameter $A$ equals 87.6.

If we now choose the ratio of the slopes between adjacent segments to be constant, and equal to two, the segmented approximation of the $A$ law is that defined by the CCITT. Table 4-1 defines the absolute values for the gradient of each slope, for this particular case.

**Table 4-1  Absolute Values for the Gradient of each Segment's Slope, Using the A Law**

| $g'(X) = \tan \alpha$ | Region of validity | Segment number $I$ |
|---|---|---|
| 16 | $0 \leq X \leq \frac{1}{64}$ | 1 |
| 8 | $\frac{1}{64} \leq X \leq \frac{1}{32}$ | 2 |
| 4 | $\frac{1}{32} \leq X \leq \frac{1}{16}$ | 3 |
| 2 | $\frac{1}{16} \leq X \leq \frac{1}{8}$ | 4 |
| 1 | $\frac{1}{8} \leq X \leq \frac{1}{4}$ | 5 |
| $\frac{1}{2}$ | $\frac{1}{4} \leq X \leq \frac{1}{2}$ | 6 |
| $\frac{1}{4}$ | $\frac{1}{2} \leq X \leq 1$ | 7 |

The objective of any companding system must be to enhance the quality of quantization for small signal values, compared to that attainable by uniform quantization. The transfer function for uniform quantization has a slope equal to unity, while that for $A$ law companding is given for the first segment by Eq. (4.31). The difference in slope is known as the *companding advantage,* or *improvement,* and for the $A$ law is equal to $A/(1 + \log A)$.

Although in this brief description we have only considered the $A$ law, it is clear that similar arguments apply to the segmentation of the $\mu$ law. The Bell T1 telephone system currently uses within their D2 multiplex, a segmented approximation of the $\mu$ law, defined by Eq. (4.29). The compression parameter $\mu$ has a value of 255, eight different quantum step sizes are employed, and there are 15 separate segments.

We have seen how smooth companding curves may be approximated as a series of chords, and have briefly described practical companders that employ this arrangement in order to reduce equipment costs. Three questions remain:

1. How many segments are required for a good approximation?
2. How many quantum steps in total should be specified?
3. Should there be an equal number of quantum steps per segment?

The answers to these questions are to be found by subjective testing, equipment cost information, and in the theoretical analysis of the expected signal-to-quantizing-noise ratio for different input signal levels. The next section deals with the last of these considerations.

## 4.4  SIGNAL-TO-QUANTIZATION-NOISE RATIO—SINUSOIDAL SIGNALS

Our aim here is to provide mathematical precision to the qualitative arguments given in the previous sections. We shall obtain expressions for the signal-to-quantizing-noise power as a function of the input signal amplitude; and use the results to plot the appropriate graphs.

The methods of nonuniform quantization that have been described, essentially aim to reduce the quantizing error by choosing the quantum step sizes in such a way that they best suit the statistical properties of the message signal. Consequently, any analysis of the type envisaged here must define a particular probability amplitude distribution for the message. We shall assume a sinusoidal message signal given by

$$x(t) = a \cos (\omega t + \theta) \qquad (4.34)$$

where $a$ represents the peak amplitude.

It is clear that the probability distribution of such a signal is unlike normal speech. However, since waveforms of this type are used extensively for test purposes it is a useful analysis to make. Later, within Sec. 4.5, we shall extend the treatment to exponential distributions as found in speech.

### 4.4.1 Uniform Quantization Case

The average signal power $S$ of a signal is by convention taken to be equivalent to the mean square value. Therefore, if the message signal $x(t)$ is given by Eq. (4.34), then

$$S = \frac{a^2}{2} \qquad (4.35)$$

The mean quantization noise power $N_Q$, for uniform quantization is given by Eq. (4.4) as

$$N_Q = \frac{(\Delta v)^2}{12}$$

Therefore the signal-to-quantization-noise power is given by

$$\frac{S}{N_Q} = \frac{a^2}{2} \frac{12}{(\Delta v)^2} = \frac{6a^2}{(\Delta v)^2} \qquad (4.36)$$

If the number of quantization steps is $n$, and the message is restricted to the normalized range $-1$ to $+1$, then the relationship between step size $\Delta v$, and $n$ is given by

$$\Delta v = \frac{2}{n} \qquad (4.37)$$

Each of the quantization steps for a binary system must be defined by a two-state coded word. If the length of the word required to specify $n$ separate levels is $b$ binary digits, then Eq. (4.37) may be rewritten as

$$\Delta v = \frac{2}{2^b - 1} \qquad (4.38)$$

Substituting Eq. (4.38) into Eq. (4.36) gives:

$$\frac{S}{N_Q} = \frac{6a^2}{4} (2^b - 1)^2 = \frac{3}{2} a^2 (2^b - 1)^2 \qquad (4.39)$$

Expressing Eq. (4.39) in terms of decibels, the result is obtained that

$$\left( \frac{S}{N_Q} \right)_{dB} = 10 \log_{10} \frac{3}{2} + 20 \log_{10} a + 20 \log_{10} (2^b - 1) \qquad (4.40)$$

The relationship between signal-to-quantizing-noise ratio and message signal amplitude is shown in Fig. 4-8, which has been plotted using Eq. (4.40), for commonly used values of $n$, or $(2^b - 1)$. The message signal is constrained to the normalized amplitude range of $\pm 1$, or 0 dB.

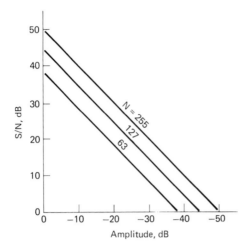

**Fig. 4-8** Signal-to-quantizing-noise ratio, plotted against signal amplitude for uniform quantization.

### 4.4.2 Nonuniform Continuous Quantization Case

The relationship between the message signal amplitude $x$, and its compressed equivalent $y$ at a particular instant in time may be generally expressed as

$$y = g(x) \tag{4.41}$$

In this case, as shown in Fig. 4-9, provided that the number of quantization steps is sufficiently large the result is obtained that

$$\frac{\Delta y}{\Delta x} = \frac{dy}{dx} = g'(x) \tag{4.42}$$

If the interval $\Delta x$ is examined over a long period of time, then the quantization noise contribution $N_{\Delta x}$ for this interval is given by Eq. (4.4) as

$$N_{\Delta x} = \frac{(\Delta x)^2}{12} \tag{4.43}$$

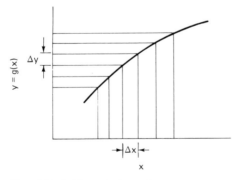

**Fig. 4-9** Differentiation of the companding curve.

Let the frequency of occurrence of each amplitude value, for the signal $x(t)$, be defined by the probability function $p(x)$. Consequently, the total quantization noise is given by

$$N_Q = \int_{a_1}^{a_2} N_{\Delta x} p(x)\, dx \tag{4.44}$$

Substituting Eq. (4.43) into Eq. (4.44) gives:

$$N_Q = \int_{a_1}^{a_2} \frac{(\Delta x)^2}{12} p(x)\, dx \tag{4.45}$$

Here the signal $x(t)$ is considered to be limited to the range defined by $a_1$ and $a_2$ where

$$-1 \le a_1 < a_2 \le 1$$

Equation (4.45) may be expressed in terms of $y$, by substituting for $\Delta x$, using Eq. (4.42). Therefore,

$$N_Q = \frac{(\Delta y)^2}{12} \int_{a_1}^{a_2} \frac{p(x)}{[g'(x)]^2}\, dx \tag{4.46}$$

Both the message signal $x(t)$ and its compressed equivalent are constrained to lie within the normalized range $-1$ to $+1$. If, then, the compressed signal is uniformly quantized using $n$ quantum steps, the size of each step is given by

$$\Delta y = \frac{2}{n} \tag{4.47}$$

Substituting Eq. (4.47) into Eq. (4.46) gives:

$$N_Q = \frac{1}{3n^2} \int_{a_1}^{a_2} \frac{p(x)}{[g'(x)]^2}\, dx \tag{4.48}$$

This important result defines the quantization noise $N_Q$ in terms of the compression $g(x)$, and the probability amplitude distribution $p(x)$. We shall now derive an expression for $p(x)$, assuming a sinusoidal message $x(t)$, of amplitude $a$, given by Eq. (4.34) as

$$x(t) = a \cos(\omega t + \theta)$$

Let us consider $t$ constant, and examine samples occurring for values of $\theta$ existing between $-\pi$ and $\pi$. Since these samples can occur with an equal probability for any value of $\theta$, the distribution $p(\theta)$ is given by

$$p(\theta) = \frac{1}{2\pi} \qquad \text{for } -\pi \le \theta \le \pi \tag{4.49}$$

Rearranging the defining Eq. (4.34), of the message signal gives:

$$\theta + \omega t = \cos^{-1} \frac{x}{a} \tag{4.50}$$

Therefore,[16]

[16] This is derived rigorously in several texts, see for example B. P. Lathi, *An Introduction to Random Signals, and Communication Theory,* International Textbook Co., Pa., 1968, p. 165.

$$p(x) = \frac{1}{\pi} \frac{1}{\sqrt{a^2 - x^2}} \qquad \text{for } -a \le x \le a \qquad (4.51)$$

Substituting Eq. (4.51) into Eq. (4.48) gives:

$$N_Q = \frac{1}{3\pi n^2} \int_{a_1}^{a_2} \frac{1}{\sqrt{a^2 - x^2} \, [g'(x)]^2} \, dx \qquad (4.52)$$

This result is valid for sinusoidal message signals given by Eq. (4.34), irrespective of the chosen compression characteristic $g(x)$. Therefore we may, by calculating $g'(x)$, determine the quantizing noise for any given companding law. It is instructive to derive expressions for both the $\mu$ and $A$ laws, as shown below.

**4.4.2.1   The $\mu$ Law.**   The defining Eq. (4.29) of the $\mu$ law may be written as

$$y = \frac{\log (1 + \mu x)}{\log (1 + \mu)} \qquad \text{for } x \ge 0$$

Therefore,

$$\frac{dy}{dx} = g'(x) = \frac{1}{\log (1 + \mu)} \frac{\mu}{(1 + \mu x)} \qquad \text{for } x \ge 0 \qquad (4.53a)$$

Similarly for negative values of $x$, the result is obtained that

$$g'(x) = \frac{1}{\log (1 + \mu)} \frac{\mu}{(1 - \mu x)} \qquad \text{for } x \le 0 \qquad (4.53b)$$

Let

$$k = \frac{\mu}{\log (1 + \mu)} \qquad (4.54)$$

Substituting Eq. (4.53) into Eq. (4.52) gives:

$$N_Q = \frac{1}{3\pi n^2} \frac{1}{k^2} \left[ \int_{-a}^{0} \frac{(1 - \mu x)^2}{\sqrt{a^2 - x^2}} \, dx + \int_{0}^{a} \frac{(1 + \mu x)^2}{\sqrt{a^2 - x^2}} \, dx \right] \qquad (4.55)$$

Integrating Eq. (4.55), and collecting terms gives:

$$N_Q = \frac{1}{3(kn)^2} \left[ 1 + \frac{(\mu a)^2}{2} + \frac{4\mu a}{\pi} \right] \qquad (4.56)$$

The signal power $S$ of the message $x(t)$ is given by Eq. (4.35) as

$$S = \frac{a^2}{2}$$

Therefore the signal-to-quantizing-noise ratio for the sinusoidal message signal $x(t)$, using the $\mu$ law, is given by

$$\frac{S}{N_Q} = \frac{3a^2}{2} (kn)^2 \left[ 1 + \frac{(\mu a)^2}{2} + \frac{4\mu a}{\pi} \right]^{-1} \qquad (4.57)$$

This result is shown graphically in Fig. 4-10, where it is compared to the curve obtained for uniform quantization; equivalent to $\mu$ equal to zero.

**4.4.2.2   The $A$ Law.**   The defining Eq. (4.30) of the $A$ law may be written as

$$y = \frac{Ax}{1 + \log A} \qquad \text{for } 0 \le |x| \le \frac{1}{A} \qquad (4.58a)$$

$$y = \frac{1 + \log (Ax)}{1 + \log A} \qquad \text{for } \frac{1}{A} \le |x| \le 1 \qquad (4.58b)$$

Therefore,

$$\frac{dy}{dx} = g'(x) = \frac{A}{1 + \log A} \qquad \text{for } 0 \le |x| \le \frac{1}{A} \qquad (4.59a)$$

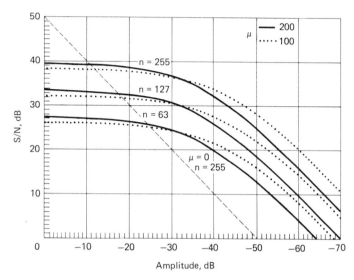

**Fig. 4-10**  Signal-to-quantizing-noise ratio, plotted against signal amplitude for continuous, nonuniform quantization ($\mu$ law).

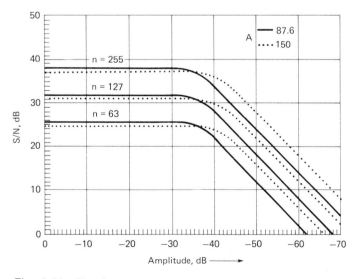

**Fig. 4-11**  Signal-to-quantizing-noise ratio, plotted against signal amplitude for continuous, nonuniform quantization (A law).

$$\frac{dy}{dx} = g'(x) = \frac{1}{(1 + \log A)} \frac{1}{|x|} \qquad \begin{array}{l} \text{for } -1 \le x < \frac{1}{A} \\ \text{and } \frac{1}{A} < x \le 1 \end{array} \qquad (4.59b)$$

Let

$$K = \frac{1}{1 + \log A}$$

Substituting Eq. (4.59) into Eq. (4.52) gives:

$$N_Q = \frac{1}{3\pi n^2} \frac{1}{K^2} \left( 2 \int_0^{1/A} \frac{1}{A^2 \sqrt{a^2 - x^2}} \, dx \; + 2 \int_{1/A}^{a} \frac{x^2}{\sqrt{a^2 - x^2}} \, dx \right) \qquad \text{for } a > \frac{1}{A} \quad (4.60a)$$

$$N_Q = \frac{1}{3\pi n^2} \frac{1}{K^2} \left( 2 \int_0^{a} \frac{1}{A^2 \sqrt{a^2 - x^2}} \, dx \right) \qquad \text{for } a \le \frac{1}{A} \quad (4.60b)$$

Integrating Eq. (4.60) and collecting terms gives:

$$N_Q = \frac{1}{3\pi (KAn)^2} \left\{ [2 - (aA)^2] \sin^{-1} \left( \frac{1}{aA} \right) + \frac{\pi}{2} (aA)^2 \right.$$

$$\left. + \sqrt{(aA)^2 - 1} \right\} \qquad \text{for } a > \frac{1}{A} \quad (4.61a)$$

$$N_Q = \frac{1}{3(KAn)^2} \qquad \text{for } a \le \frac{1}{A} \qquad (4.61b)$$

The signal power $S$ of the message signal $x(t)$ is given by Eq. (4.35) as

$$S = \frac{a^2}{2}$$

Therefore the signal-to-quantizing-noise ratio for the sinusoidal message signal $x(t)$, using the $A$ law, is given by

$$\frac{S}{N_Q} = \frac{3\pi (aKAn)^2}{2} \left\{ [2 - (aA^2)] \sin^{-1} \frac{1}{aA} + \frac{\pi}{2} (aA)^2 \right.$$

$$\left. + \sqrt{(aA)^2 - 1} \right\}^{-1} \qquad \text{for } a > \frac{1}{A} \quad (4.62a)$$

$$\frac{S}{N_Q} = \frac{3(aKAn)^2}{2} \qquad \text{for } a \le \frac{1}{A} \qquad (4.62b)$$

This result is shown graphically in Fig. 4-11, where it is compared to the curve obtained for uniform quantization.

In Sec. 4.2 we noted that the normal design criterion of a nonuniformly quantized system is that the signal-to-quantizing-noise ratio should always exceed some acceptable figure (e.g., 35 dB), over the greatest possible range of signal amplitudes (e.g., 30 dB). If we now examine Figs. 4-10 and 4-11, it is clear that a signal-to-noise ratio of better than 35 dB can be achieved, over uniform quantization using the $\mu$ law, or $A$ law companding; however, the possible range of signal amplitudes differs significantly. The amplitude ranges, for 35 dB signal-to-noise, and 255 quantum steps $n$, are

$$\text{Uniform quantization} \qquad = 16 \text{ dB}$$
$$\mu \text{ law companding } (\mu = 100) \; = 38 \text{ dB}$$
$$A \text{ law companding } (A = 87.6) = 40 \text{ dB}$$

These figures are clear evidence that the useful signal amplitude range can be greatly extended by adopting companding techniques.

### 4.4.3 Segmented Quantizing

We may imagine, for the purpose of this analysis, a sinusoidal signal of amplitude sufficiently large that several segments are traversed. In this case, each of the traversed segments contributes to the total quantizing noise $N_Q$. We shall calculate the quantizing noise for sinusoidal signals of increasing amplitude, such that initially the signal is constrained to lie within the first segment $N_{Q1}$, then the first and second $N_{Q2}$, and finally the first, second, and third segments. This will enable us to generalize, and determine the quantizing noise for a sinusoid that traverses all segments, up to and including the $i$th. We shall assume in this analysis segmented $A$ law companding, with the compression parameter $A$ set at 87.6, and the limits of each segment as defined by Table 4-1.

The quantizing noise due to each segment, for sinusoidal signals that are symmetrical about the origin, may be derived from Eq. (4.52) as

$$N_Q = \frac{1}{3\pi n^2} \int_{a_1}^{a_2} \frac{2}{\sqrt{a^2 - x^2} \, [g'(x)]^2} \, dx \tag{4.63}$$

Here the limits of integration are given by the extent of the segment, as defined in Table 4-1. Conveniently, $g'(x)$ remains constant throughout each segment, and may be represented by the slope of the companding curve ($\tan \alpha$). Therefore, the quantizing noise of the first segment $N_{Q1}$ is given by

$$N_{Q1} = \frac{1}{3\pi n^2} \int_0^a \frac{2}{\sqrt{a^2 - x^2} \, (\tan \alpha_1)^2} \, dx \tag{4.64}$$

$$N_{Q1} = \frac{1}{3n^2} \frac{1}{2^8} F(x)_1 \tag{4.65}$$

where $F(x)_i = \dfrac{\sin^{-1}(x/a)}{\pi/2}$ within the $i$th segment.

However, for the first segment ($i = 1$): $a = x$ and $F(x)_1 = 1$. Therefore,

$$N_{Q1} = \frac{1}{3n^2} \frac{1}{2^8} \tag{4.66}$$

The quantizing noise for a signal traversing the first and second segments $N_{Q2}$, using the limits of integration shown in Table 4-1, is given by

$$N_{Q2} = \frac{1}{3n^2} \frac{1}{2^8} F(\tfrac{1}{64}) + \frac{1}{3n^2} \frac{1}{2^6} [F(x)_1 - F(\tfrac{1}{64})] \tag{4.67}$$

$$N_{Q2} = \frac{1}{3n^2} \frac{1}{2^6} - \frac{1}{n^2} \frac{1}{2^8} F(\tfrac{1}{64}) \tag{4.68}$$

Similarly, the quantizing noise for a signal traversing the first, second, and third segments ($N_{Q3}$) is given by

$$N_{Q3} = \frac{1}{3n^2} \frac{1}{2^6} F(\tfrac{1}{32}) - \frac{1}{n^2} \frac{1}{2^8} F(\tfrac{1}{64}) + \frac{1}{3n^2} \frac{1}{2^4} [F(x)_1 - F(\tfrac{1}{32})] \quad (4.69) \qquad (4.69)$$

$$N_{Q3} = \frac{1}{3n^2} \left[ \frac{1}{2^4} - \frac{3}{2^6} F(\tfrac{1}{32}) - \frac{3}{2^8} F(\tfrac{1}{64}) \right] \qquad (4.70)$$

We may generalize from Eqs. (4.66), (4.68), and (4.70) that the quantizing noise $N_Q$ that occurs for a sinusoidal signal which traverses all segments up to and including the $i$th segment is given by

$$N_Q = \frac{1}{3n^2} \left[ \frac{1}{(2^{5-i})^2} - 3 \sum_{j=1}^{i-1} \frac{F(1/2^{7-j})}{(2^{5-j})^2} \right] \qquad (4.71)$$

Once again the signal power $S$ of the message signal $x(t)$ is given by Eq. (4.35) as

$$S = \frac{a^2}{2}$$

Therefore the signal-to-quantizing-noise ratio for the sinusoidal message signal $x(t)$, using the segmented $A$ law, is given by

$$\frac{S}{N_Q} = \frac{3(an)^2}{2} \left[ \frac{1}{(2^{5-i})^2} - 3 \sum_{j=1}^{i-1} \frac{F(1/2^{7-j})}{(2^{5-j})^2} \right]^{-1} \qquad (4.72)$$

This result is valid for any segment, and is shown graphically in Fig. 4-12, where it is compared to the curves obtained for uniform quantization, and continuous $\mu$ law companding. It is interesting to note that $F$ is a function of both the amplitude and wavelength of the message signal $x(t)$. We shall now describe the curve traced by Eq. (4.72) qualitatively.

The first segment is perhaps the most important, since it is the small response that is improved by companding. A comparison between the curves pertaining to

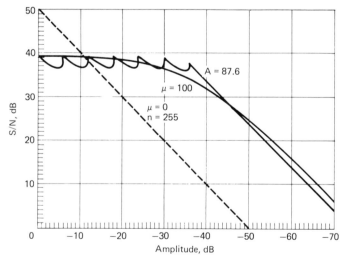

**Fig. 4-12** Signal-to-quantizing noise ratio, plotted against signal amplitude for segmented (A law); continuous nonuniform ($\mu$ law); and uniform quantization.

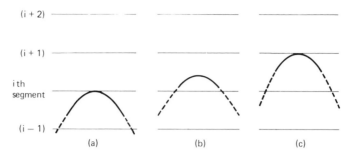

**Fig. 4-13**   The effect of increasing amplitude on quantizing error.

uniform quantization and segmented companding shows that at small amplitudes there is a difference between them of approximately 25 dB.   This quantity is referred to as the *companding advantage* as described earlier.

The second and higher-order segments occur during the logarithmic portion of the $A$ law characteristic.   Here the absolute position of the peak amplitude within the superior segment exerts a major influence on the signal-to-quantizing-noise ratio. Intuitively we may explain this effect by noting that the quantizing error of a sinusoidal signal is concentrated at the amplitude peaks, where the number of quantum steps intersected is small.   Let us consider then the peak or noisy portion of the sinusoid (solid line) moving up within a segment, as shown in Fig. 4-13.

Figure 4-13a defines the situation where the noisy portion of the sinusoid occurs mainly in the lower segment $i - 1$.   Consequently, the major contribution to the total quantizing noise given by Eq. (4.71) is due to this segment.   The signal-to-noise ratio is at a maximum for this case.

If the signal level is now gradually increased, the noisy portion of the sinusoid effects both the central $i$, and lower $i - 1$ segments.   In this case although the signal power increases a little, the noise contribution, now due to two segments, increases by a greater amount, and the signal-to-noise ratio gradually drops in consequence.

Eventually a point is reached, as shown in Fig. 4-13b where the noisy portion is almost completely contained within the central segment $i$.   When this occurs the noise contribution is mainly due to this single segment only, and consequently the total quantization noise reduces.

Further increases in the signal level result in an increase in the signal power, while the noise power remains constant, or reduces slightly.   Thus the signal-to-noise ratio improves until the situation depicted by Fig. 4-13c is arrived at, whereupon the cycle repeats.

Figure 4-12 shows that for a sinusoidal message waveform, there is a periodic degradation in the signal-to-noise ratio of the type described.   This effect rules out the possibility of using single frequency sinusoidal signals for the measurements of quantization noise within segmented companding systems, because any slight level change could result in a measurement discrepancy of up to 3 dB.   This problem may be overcome by using band-limited noise as a test signal.

## 4.5   QUANTIZATION NOISE FOR A SPEECH SIGNAL

For the sake of completeness we shall now return to Eq. (4.48), which defines the quantization noise $N_Q$ in terms of the compression $g(x)$ and the probability amplitude distribution $p(x)$, as

$$N_Q = \frac{1}{3n^2} \int_{a_1}^{a_2} \frac{p(x)}{[g'(x)]^2} \, dx$$

For speech signals, the probability distribution $p(x)$ is exponential and has been shown[17] to be approximately given by

$$p(x) = \frac{1}{\sqrt{2} \, R} \exp \left( \frac{-\sqrt{2} \, |x|}{R} \right) \tag{4.73}$$

where $R$ = the effective root-mean-square (rms) value of the signal.

Let us assume for the purpose of this analysis, continuous $\mu$ law companding. Therefore, from Eq. (4.53) and Eq. (4.54) we get:

$$g'(x) = K \left( \frac{1}{1 + \mu_x} \right) \qquad \text{for } x \geq 0 \tag{4.74a}$$

$$g'(x) = K \left( \frac{1}{1 - \mu_x} \right) \qquad \text{for } x \leq 0 \tag{4.74b}$$

Substituting Eq. (4.73) and Eq. (4.74) into Eq. (4.48) gives:

$$N_Q = \frac{1}{3n^2} \frac{1}{K^2} \frac{1}{\sqrt{2} \, R} \left[ \int_{-1}^{0} (1 - \mu x)^2 \exp \left( \frac{\sqrt{2} \, |x|}{R} \right) dx \right.$$
$$\left. + \int_{0}^{1} (1 + \mu x)^2 \exp \left( \frac{-\sqrt{2} \, |x|}{R} \right) dx \right] \tag{4.75}$$

Strictly, the limits of integration must be $\pm 1$, however, if the probability of seeing an amplitude $p(x)$ outside the normalized range is negligible, then the limits $\pm \infty$ may be conveniently used. If this criterion is invalid, then the effect of peak clipping must also be taken into account. We shall consider this possibility within the next section; however, here we use $\pm \infty$ as the integration limits and obtain from Eq. (4.75) the result that

$$N_Q = \frac{1 + \mu^2 R^2 + \mu \sqrt{2} R}{3(nK)^2} \tag{4.76}$$

It is interesting to compare the results obtained using continuous $\mu$ law companding, for a probability distribution $p(x)$ relating to normal speech [Eq. (4.76)], and a sinusoidal signal given by Eq. (4.56), which is repeated below:

$$N_Q = \frac{1 + (\mu a)^2/2 + 4\mu a/\pi}{3(nK)^2}$$

The difference between the two results is seen to be within the last term only (when $R^2 = a^2/2$).

## 4.6  QUANTIZING IMPERFECTIONS

We have now analyzed several types of systems, in both qualitative and quantitative terms. In this section we shall pause for a moment in order to reappraise our analysis and identify the various assumptions that have been made.

[17] Davenport, "Speech-Wave Probability Distributions."

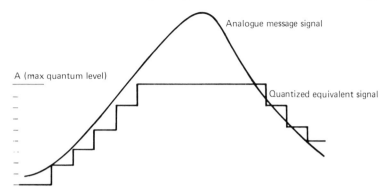

**Fig. 4-14**  Overloading due to peak clipping.

### 4.6.1  Peak Clipping

Ideal quantization systems are arranged such that the amplitude peaks of the message signal are just contained within the quantizing range. In this case the operating range of the equipment must be impractically large in order to cater for very occasional high-level signals. We may compromise, however, by specifying a quantizing range suited to the most probable amplitudes only, and ignoring the infrequent peaks. This results in *peak clipping,* as is shown in Fig. 4-14.

Previously we have assumed that the message signal is constrained to lie within the range ±1, and that the probability of seeing an amplitude outside this range is zero. We shall now consider that an occasional sample falls outside the permitted range, and that the frequency of occurrence is about 1 in $10^3$ samples. In this case, we see from Eq. (4.73) that

$$\exp\left(\frac{-\sqrt{2}\cdot|\pm1|}{R}\right) = 10^{-3} \quad \simeq \exp(-7) \tag{4.77}$$

Therefore,

$$\frac{\sqrt{2}}{R}1 \simeq 7 \tag{4.78}$$

In dBs

$$[R^2]_{dB} \simeq -14 \text{ dB} \tag{4.79}$$

This result indicates that the outcomes obtained for modulation with a voice signal are valid provided that the mean signal power is less than −14 dB.

Listening tests have confirmed that slight peak clipping does not impair the quality of speech signals significantly. The error due to peak clipping is just perceptible,[18] for speech signals, above about −9 dB. However, most practical systems allow about 14 dB below the rms signal level before clipping occurs. This is imperceptible.

The mean square error, or noise power $N_c$ due to peak clipping, assuming that the permitted range extends from −1 to +1, is given by

---

[18] D. L. Richards, "Transmission Performance of Telephone Networks, Containing PCM Links," *Electronics Record,* vol. 118, 1968, p. 1245.

$$N_c = 2 \int_1^\infty (x-1)^2 p(x)\, dx \tag{4.80}$$

Substituting Eq. (4.73) into Eq. (4.80) gives:

$$N_c = \frac{2}{\sqrt{2}\, R} \int_1^\infty (x-1)^2 \exp\left(\frac{-\sqrt{2}\,|x|}{R}\right) dx \tag{4.81}$$

Let

$$u = (x-1)^2$$

Therefore,

$$du = 2(x-1)\, dx$$

Let

$$dv = \exp\left(\frac{-\sqrt{2}\,|x|}{R}\right) dx$$

Let

$$K = \frac{\sqrt{2}}{R}$$

Therefore,

$$v = \frac{\exp(-Kx)}{-K} \tag{4.82}$$

Substituting the above terms into Eq. (4.81) gives the result:

$$N_c = K \int_1^\infty u\, dv$$

$$= K\left(uv - \int_1^\infty v\, du\right)$$

$$= K\left[\frac{(x-1)^2 \exp(-Kx)}{-K}\right]_1^\infty - K \int_1^\infty \left[\frac{\exp(-Kx)}{-K}\right] 2(x-1)\, dx$$

$$= 0 + 2\int_1^\infty (x-1)\exp(-Kx)\, dx \tag{4.83}$$

Let

$$m = x - 1$$

Therefore,

$$dm = dx$$

Let

$$dn = \exp(-Kx)\, dx$$

Therefore,

$$n = \frac{\exp(-Kx)}{-K}$$

Substituting the above terms into Eq. (4.83) gives the result:

$$N_c = 2\int_1^\infty m\, dn = 2\left(mn - \int_1^\infty n - dm\right)$$

$$= 2\left[\frac{(x-1)\exp(-Kx)}{-K}\right]_1^\infty - 2\int_1^\infty \frac{\exp(-Kx)}{-K}\, dx$$

$$= 0 + \frac{2}{K}\int_1^\infty \exp(-Kx)\, dx$$

$$= \frac{2}{K}\left[\frac{\exp(-Kx)}{-K}\right]_1^\infty$$

$$= \frac{2}{K^2}\exp(-K) \tag{4.84}$$

Substituting Eq. (4.82) into Eq. (4.84) gives:

$$N_c = R^2 \exp\left(\frac{-\sqrt{2}}{R}\right) \tag{4.85}$$

The total noise due to the process of quantizing $N_T$ is given by the sum of the clipping noise $N_c$, and the quantizing error $N_Q$ calculated earlier. Therefore, for uniform quantization we obtain from Eq. (4.4):

$$\begin{aligned}
N_T &= N_Q + N_c \\
&= \frac{(\Delta v)^2}{12} + R^2 \exp\left(-\sqrt{2}R\right)
\end{aligned} \tag{4.86}$$

Substituting Eq. (4.37) into Eq. (4.86) gives:

$$N_T = \frac{2}{3n^2} + R^2 \exp\left(-\sqrt{2}R\right) \tag{4.87}$$

where $n =$ the number of quantization steps, in the range $-1$ to $+1$.

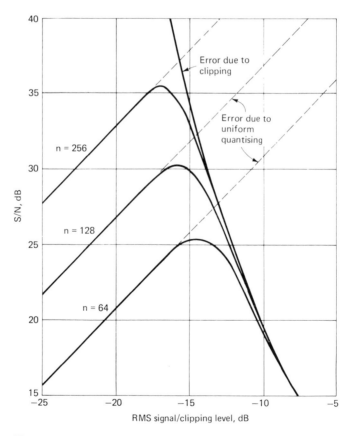

**Fig. 4-15**  The effect of peak clipping on signal-to-quantizing-noise ratio. (From Kenneth William Cattermole, *Principles of Pulse Code Modulation*, Iliffe, London, 1969.)

Now, by definition the input signal power $S$ equals $R^2$. Therefore the signal-to-total-noise ratio is given by

$$\frac{S}{N_T} = \frac{3}{2} n^2 R^2 + \exp(\sqrt{2}R) \tag{4.88}$$

Expressing Eq. (4.88) in terms of decibels, the result is obtained that

$$\left[\frac{S}{N_T}\right]_{dB} = 10 \log_{10}\left[\frac{3n^2}{2} R^2 + \exp(\sqrt{2}R)\right] \tag{4.89}$$

Consequently if the effective rms value $R$ is small in value, the signal-to-noise ratio will approach the result obtained in Eq. (4.40); that is the effect of clipping is negligible. However if $R$ is large in value, the major noise contribution is due to clipping. Equation (4.89) has been plotted in Fig. 4-15, for three different values of $n$.

### 4.6.2  Idle Channel Noise

We have assumed throughout this chapter that a message signal of varying amplitude is present at the quantizer input. This criterion is only approximated for signals of extremely small amplitude level, and in the limit when the message signal may be considered no longer present the previous analyses are invalid.

Random fluctuations, possibly of magnitude less than one quantum step $\Delta x$, and the position of the quiescent level in relation to the quantizing scale, determine the system perturbations for the no signal condition, that is, idle channel noise. In this case the mean error calculations become meaningless since one requires prior knowledge of the exact form of the message signal.

Suppose that the quiescent level for the no signal condition is positioned exactly midway between two quantization thresholds, as shown in Fig. 4-16. Provided the random fluctuations do not exceed $\pm\Delta x/2$, no threshold is crossed, and no perturbations are introduced. The idle channel noise is zero.

On the other hand, suppose that the quiescent level coincides, or lies close to a threshold. In this case, the smallest fluctuation will cause a new quantum level to be registered; and produces a square wave of amplitude $\Delta x/2$ at the output to the quantizer, as shown in Fig. 4-16.

The maximum idle channel noise $N_i$ power is, by definition, equivalent to the square of the output amplitude, which for an equiprobable periodic square wave gives:

$$N_i = \frac{(\Delta x)^2}{4} \tag{4.90}$$

This result suggests that the noise power attributed to the no-signal condition is greater than when a signal is present, which is given by $(\Delta x)^2/12$. However, although this is possible, one does not normally record such a high result. We shall return to Fig. 4-16 in order to consider the problem in more depth.

Let the probability of overtaking the threshold be $p$. Therefore, the probability of not overtaking the threshold is $1 - p$. The output noise power $N_i$ may then generally be expressed as

$$N_i = p(1 - p)(\Delta x)^2 \tag{4.91}$$

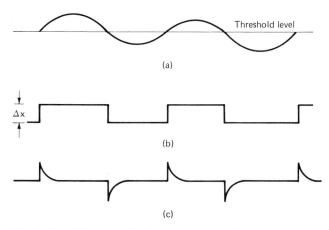

(a)

(b)

(c)

**Fig. 4-16**   Idle channel noise: (a) quantizer input, with quiescent level near threshold; (b) decoded output signal; (c) final output, after filtration (typical oscilloscope trace).

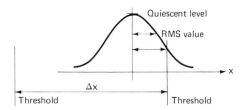

**Fig. 4-17**   Amplitude distribution of fluctuations about the quiescent level.

The probability $p$ is dependent on two factors:

1. The separation $d$ between the quiescent level and the nearest threshold.
2. The amplitude distribution of the fluctuations about the quiescent level. Here we shall assume a gaussian distribution, of rms amplitude $a$, as shown in Fig. 4-17.

For a gaussian distribution $p(x)$ is given by

$$p(x) = \frac{1}{a\sqrt{2\pi}} \exp\left(\frac{-1}{2}\frac{x^2}{a^2}\right)$$   (4.92)

We are only interested in the probability $p$ of crossing the nearest threshold. (Note that we limit ourselves to considering fluctuations that cross one threshold only.) Therefore $p$ is given by

$$p = \frac{1}{a\sqrt{2\pi}} \exp\left(\frac{-1}{2}\frac{d^2}{a^2}\right)$$   (4.93)

The idle channel noise power has been plotted in Fig. 4-18, as a function of the quiescent level position, using Eqs. (4.91) and (4.93). Clearly if the quiescent level has a tendency to drift, the noise power that occurs will vary in magnitude between a very low value and $(\Delta x)^2/4$. However, in practice the quiescent level remains fairly constant in value and can sustain particular noise power values for long periods.

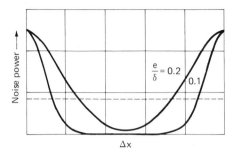

**Fig. 4-18**  Idle channel noise as a function of the quiescent level. (From Kenneth William Cattermole, *Principles of Pulse Code Modulation,* Iliffe, London, 1969.)

The figure clearly indicates that the most frequent noise levels are considerably less than that defined by Eq. (4.90).

In the foregoing the noise power has been examined for the no signal condition only. Normally, however, there will always be some low-frequency signal present, say 50 or 60 Hz. This is referred to as *mains hum* and is due to crosstalk effects between the input and the power supply.

The effect of hum may be reduced by passing the quantizer output signal through a high-pass filter, such that the fundamental frequency of the hum is strongly attenuated. The harmonics will nevertheless traverse the filter, as shown in Fig. 4-16c, and will therefore contribute to the idle channel noise.

Let the time constant of the high-pass filter $= \frac{1}{2}\pi f$. Each output spike, of unit amplitude, is given by $\exp(-2\pi ft)$.

If we consider a single spike in isolation and evaluate it over an infinite period of time we find that the energy is proportional to

$$\int_{0}^{\infty} [\exp(-2\pi ft)]^2 \, dt = \frac{1}{4\pi f} \tag{4.94}$$

Since the hum frequency $F$ is small compared to $f$, then the spikes may be considered separate from one another, and the mean square noise $N_i$ is approximately given by

$$N_i \simeq (\Delta x)^2 \left(\frac{F}{2\pi f}\right) \tag{4.95}$$

The important point to note about this result is that the normal noise level given by $(\Delta v)^2/12$ is higher than that given by Eq. (4.95), provided $f/F$ is less than $12/2\pi$.

In conclusion we should consider the effect of crosstalk from one or more adjacent message channels. This introduces random fluctuations about the quiescent level, and analysis is similar to the no signal condition. Consequently the noise power in the worst case could reach a value of $(\Delta v)^2/4$, but normally is much lower than this.

### 4.6.3  Practical Limitations

We have assumed throughout this chapter that each quantizing step has a precisely defined position on either a uniform, or a nonuniform scale, and that the origin of this scale is coincident with the quiescent amplitude of the message signal. In practice, due to equipment tolerances this assumption is invalid and requires further consideration. We must consider both random and systematic departures from the ideal.

The random departures are likely to be due to individual component inaccuracies, which are in turn affected by temperature and supply voltage variations. Each quantum step may be defined by a separate circuit element. In this case variations in the signal amplitude will cause random inaccuracies in the quantized equivalent signal, since the ideal quantizing scale can only be realized if each circuit element is exactly correct. Further inaccuracies may occur due to the impossibility of defining a threshold with an infinitesimal width; however, due to modern high-speed switching circuits the effect is minimal.

The investigation of these random effects is somewhat complex, and should only be undertaken with a particular equipment and message signal in mind. An analysis of this type has been performed for the $\mu$ law coder and is described in a paper by Mann, Straube, and Villars.[19]

Systematic departures from the ideal are normally due to the introduction of a dc bias within the quantizing system. We shall consider first the case of the quiescent level of the message signal being displaced away from the center of the quantizing scale. In a uniformly quantized system such a displacent will have no effect, provided peak limiting does not occur. The quantizing noise power will always be specified by Eq. (4.4).

Since nonuniform quantizing systems are designed to quantize small amplitude signals more finely than the larger ones, it follows that the quiescent level must lie on the center of symmetry of the companding curve. A displacent of the type envisaged will result in the more numerous small amplitude signals being quantized using a quantum interval which is larger than specified. Consequently the signal-to-noise ratio for small amplitudes will be lower than predicted.

Companding laws that exhibit a logarithmic dependence throughout the quantizing range, tend, in theory at least,[20] to be highly sensitive to the proper positioning of the quiescent level. However, the measured performance of practical PCM systems using diode companders has suffered only a 2-dB impairment of signal-to-noise ratio, while theory predicted 6 dB for the same displacement. This has been attributed[21] to the linear, rather than logarithmic characteristic of the diodes at small signal amplitudes. Companding laws that exhibit a linear central section approach uniform quantization during this portion of the characteristic. Such laws are relatively insensitive to small displacements.

Finally we should consider, as shown in Fig. 4-19, the effect of introducing a dc

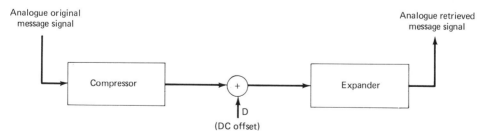

**Fig. 4-19**  Introducing a dc bias after the compressor.

[19] H. Mann, H. M. Straube, and C. P. Villars, "A Companded Coder System for an Experimental PCM Terminal," *Bell System Technical Journal*, vol. 41, 1962, pp. 173–226.
[20] Smith, "Instantaneous Companding."
[21] Shennum and Gray, "Performance Limitations of a Practical PCM Terminal."

bias after the compressor rather than before it. In this case the center of symmetry for the compressor and expander will be different.

Ideally the compression $g(x)$, and expansion $g^{-1}(y)$ characteristics are exactly complementary, such that the expander output $x_2$ equals the compressor input $x_1$. In other words:

$$x_1 \rightarrow g(x_1) \rightarrow y_1 \rightarrow y_2 \rightarrow g^{-1}(y_2) \rightarrow x_2 \qquad (4.96a)$$

In the limit

$$x_1 = g(x_1) = y_1 = y_2 = g^{-1}(y_2) = x_2 \qquad (4.96b)$$

The effect of adding a displacent $D$ will be to make $y_1$ unequal to $y_2$. In this case $x_1$ and $x_2$ are also unequal as shown below.

$$x_1 = g(x_1) = y_1 \qquad (4.97)$$

$$y_1 + D = y_2 = g^{-1}(y_2) = x_2 \qquad (4.98)$$

Therefore,

$$x_2 = g^{-1}(y_1 + D) \qquad (4.99)$$

We may consider, as an example, that $g(n)$ is defined by the $A$ law characteristic such that

$$y = AKx = g(x) \qquad \text{for } 0 \le |x| \le \frac{1}{A} \qquad (4.100a)$$

$$y = K[1 + \log (Ax)] = g(x) \qquad \text{for } \frac{1}{A} \le |x| \le 1 \qquad (4.100b)$$

where $K = 1/(1 + \log A)$.

Therefore, for the linear portion of the characteristic, substituting Eq. (4.100a) into (4.99) gives:

$$x_2 = \frac{1}{AK}(AKx_1 + D) = x_1 + \frac{D}{AK} \qquad \text{for } 0 \le |x| \le \frac{1}{A} \qquad (4.101)$$

For the logarithmic portion of the characteristic we note from Eq. (4.98) that

$$y_2 = y_1 + D \qquad (4.102)$$

Substituting Eq. (4.100b) into (4.102) gives:

$$k(1 + \log Ax_2) = k(1 + \log Ax_1) + D \qquad \text{for } \frac{1}{A} < |x| < 1 \qquad (4.103)$$

Therefore,

$$x_2 = x_1 \exp \left( \pm \frac{D}{K} \right) \qquad \text{for } \frac{1}{A} < |x| < 1 \qquad (4.104)$$

The companding law must be skew symmetrical about the origin. Consequently, if positive amplitudes are multiplied by $\exp (+D/K)$, then negative ones are multiplied by $\exp (-D/K)$.

Figure 4-20 shows $x_1$ plotted against $x_2$ for a displaced $A$ law compander. The plot has been made based on Eqs. (4.101) and (4.104). It is interesting to

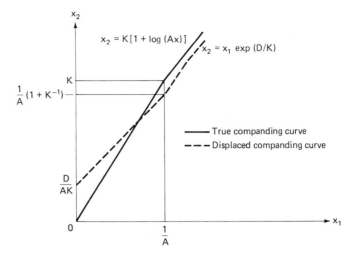

**Fig. 4-20**   The displaced A law companding curve.

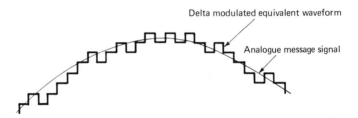

**Fig. 4-21**   Quantization using delta modulation.

note that there is a smooth, although almost linear, transition between the two main segments.

The physical outcome of adding a dc bias after the compressor is that harmonic distortion is introduced within the output signal.   For speech the effect is minimal; however, if the channel is used for conveying voice-frequency tones (data messages) then there is a problem.   Consequently, practical equipments need to give special consideration to this question.

## 4.7   OTHER FORMS OF QUANTIZING

So far in this chapter we have considered each quantum value to be referred to some fixed voltage; for example earth potential.   This is classical PCM.

We shall now investigate alternative techniques where changes from the previous quantum level, rather than absolute values, are communicated.   This is differential PCM.   For the particular case, where the maximum change that may occur at a time is one quantum level, the technique is referred to as *delta modulation,* and is illustrated in Fig. 4-21.

### 4.7.1   Delta Modulation

Essentially, in delta modulation the shape of the message signal is communicated by informing that either a positive or a negative excursion has occurred since the last sampling instant.   If the message happens to remain at a fixed dc level for a period of time this is indicated by alternating positive and negative pulses as shown in Fig. 4-21.

Since there is no limit to the number of consecutive pulses of the same sign that may occur, delta modulation systems are capable of tracking signals of any amplitude. That is, in theory at least, there is no peak clipping.

The theoretical limitation of a delta modulation system is that the message signal must not change too rapidly.   The problem, referred to as *slope clipping*, is illustrated in Fig. 4-22.   Here, although each sampling instant indicates a positive excursion, the message signal is rising too quickly, and the quantizer is unable to keep pace. We may reduce the effect of slope clipping by

1. Increasing the sampling rate
2. Increasing the quantum step size ($\Delta x$)

Subjective tests have shown[22] that in order to obtain the same transmission quality as classical PCM, very high sampling rates are required; typically 10 times the Nyquist rate.   The value of the step size $\Delta x$ is dependent on the quantizing error that can be tolerated.   This may be analyzed in a similar way to classical PCM.

Delta modulation was first invented, like classical PCM, at the ITT Laboratories, France, by the director of the laboratories E. M. Deloraine, and S. Van Mierlo and B. Derjavitch.   The first papers to be published on the subject were due to engineers of the Philips Laboratories, Holland; F. de Jager[23] and H. Van de Weg.[24]   The general invention of differential quantizing was patented at about the same time by C. C. Cutter[25] of the Bell Telephone Laboratories.

A block schematic identifying the main circuit elements within a delta modulation system is shown in Fig. 4-23.   The physical realization is very much simpler than a classical PCM quantizer of comparable quality, and is to be preferred if only one

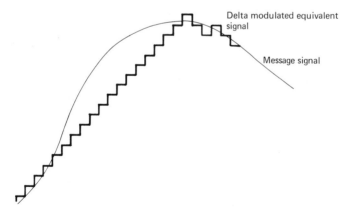

**Fig. 4-22**   Overloading due to slope clipping in delta modulation.

[22] L. Zetterberg, "A Comparison between Delta and Pulse Code Modulation," *Ericsson Technics,* vol. 2, 1955, pp. 95–154.

[23] F. de Jager, "Delta Modulation: A Method of PCM Transmission Using the One Unit Code," *Philips Research Reports,* vol. 7, 1952, pp. 542–566.

[24] H. Van de Weg, "Quantizing Noise of a Single Integration Delta Modulation System with an N-Digit Code," *Philips Research Reports,* vol. 8, 1953, pp. 367–385.

[25] U.S. Patent no. 2605361, July 29, 1952.

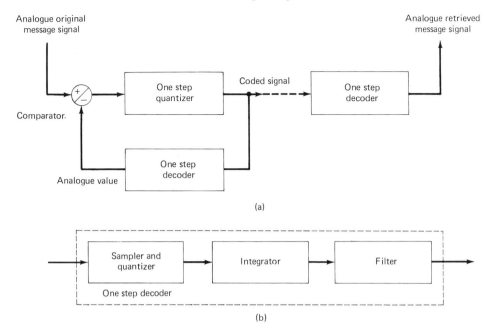

**Fig. 4-23**  The implementation of differential quantization.

message signal is involved.  However, in telephony many channels are likely to travel the same route, and may therefore be multiplexed together.  A major advantage of classical PCM is that the cost of a single quantizing unit may be shared between several channels.  This is not possible in a differential PCM system, which must store the preceding sample value of each channel.

We may summarize the main disadvantages of delta modulation when compared to classical PCM as

1. A higher sampling rate is required
2. Increased cost for multiplexed systems, since a separate quantizer for each channel is required

These drawbacks have tended to limit the development of delta modulation systems until quite recently.  However, advances in the area of integrated circuit (IC) design may change this.  It is now possible to fit a complete coder, using delta modulation, within a single IC package at low cost.  This possibility has revived interest in delta modulation for purposes where few channels are required, and a high sampling rate is permissible.  An example where these conditions exist may be found in telephony at the local network level.

We noted in Chap. 2 that the cost of a telephone network is concentrated within the local area.  Each telephone subscriber requires a separate cable link to the nearest exchange!  Deltamodulation equipment, together with cheap, simple multiplexing units, may in future be used to combine 2 or 3 telephone channels for connection to the local exchange.

### 4.7.2  Differential PCM

Differential encoding using several digits, as opposed to simple delta, has been proposed on a number of occasions for the transmission of speech and television signals. It is appropriate therefore to briefly consider the merits of this approach.

We have already seen for the case of classical PCM that the properties of the signal determine the quality, in terms of signal-to-noise ratio, for the system. This is also true for differential PCM.

Differential techniques are suited only to signals that cannot change very rapidly.

The maximum acceptable slope before the onset of slope clipping is in fact a major system parameter. However, since many practical signals such as speech (see Fig. 2-2) have a spectrum that decreases at the higher frequencies, the effect of slope clipping need not be too serious.

Clearly the onset of slope clipping is dependent on both the maximum frequency of the message signal, and on the sampling rate. If the message is never allowed to change significantly between successive samples there will be a high degree of correlation between the sample values. We may expect then, that if a particular message signal shows on average a good sample-to-sample correlation there will be an advantage in using differential, rather than classical PCM.

A theoretical study by McDonald,[26] which was based on computer analysis of recorded speech, has shown that for a sampling rate of 9.6 kHz there is very good correlation between samples. Using this model McDonald has shown a theoretical improvement in the signal-to-noise ratio of 7.14 dB for differential rather than classical PCM, with an equal number of digits. If the sampling rate is lowered to 8 kHz the degree of correlation between samples reduces and the signal-to-noise improvement drops to 5.7 dB. This improvement is roughly equivalent to adopting classical PCM, and increasing the number of digits by one.

Subjective comparisons between differential and classical PCM have confirmed the theoretical improvements identified above, for some speech samples. However, other speech samples have shown no improvement at all. This is most likely due to the greater dependence of differential methods on the properties of the signal than with classical PCM.

In conclusion differential PCM seems to offer little or no theoretical advantages compared to classical PCM. Moreover, differential quantizing circuits offer no reduction in complexity, and may not be shared between several channels.

### 4.7.3  Predictive Quantization

The differential techniques described above are in many ways incompatible with the principles of information theory. We have noted the need for a high degree of correlation between successive samples; in this case it is clear that each sample value will contain much redundant information. Essentially predictive quantization aims to remove most of the redundancy by guessing, using previous knowledge of the signal, the most likely value of the next samples, and communicating only differences between the guessed and actual values. The technique is illustrated in Fig. 4-24.

Obviously the problem of predicting sample values is eased if the basic characteristics of the message signal are known. For example, in the case of television signals the regularized format of the video information could be taken into account within the prediction process. Alternatively, in the case of speech the differences that exist between voiced and unvoiced sounds can be taken into account within the prediction process, as is discussed below.

[26] R. A. McDonald, "Signal to Noise, and Idle Channel Performance of Differential PCM Systems," *Bell System Technical Journal,* vol. 45, 1966, pp. 1123–1151.

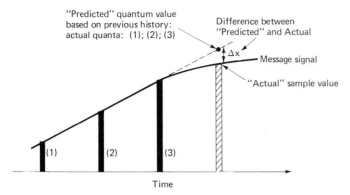

**Fig. 4-24** The technique of predictive quantization.

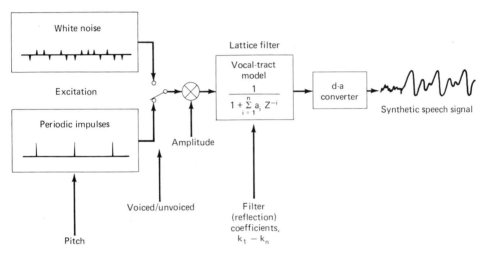

**Fig. 4-25** Linear predictive coding and speech synthesis. (From Richard Wiggins and Larry Brantingham, "Three-Chip System Synthesizes Human Speech," *Electronics,* vol. 51, no. 18, 1978. Copyright McGraw-Hill, Inc., 1978. All rights reserved.)

It should be pointed out that there are two main applications of linear predictive coding (LPC) to speech messages. First, real-time LPC systems, otherwise known as *vocoders* (voice-coders), offer the capability of much reduced data transmission bandwidths compared to conventional PCM systems. Second, the ability to have spoken words stored in reasonably sized semiconductor memories has opened the doors to a new era in the field of electronics: speech synthesis. Let us discuss the technique of speech synthesis, which has been largely pioneered by Richard Wiggins, and Larry Brantingham of Texas Instruments.[27]

Voiced sounds, such as "I," "O," and "M" have a definite pitch and can be represented by periodic impulses, of low frequency, but high amplitude. By contrast, unvoiced sounds such as "S," "F," and "SH" have a less precise nature, and may be represented by low-amplitude, random high-frequency signals, being similar to white noise.

[27] R. Wiggins and L. Brantingham, "Three-Chip System Synthesizes Human Speech," *Electronics,* vol. 51, no. 18, 1978.

In the human generation of speech, voiced and unvoiced excitations are produced by the vocal chords due to the passage of air from the lungs. The vocal tract, formed by the tongue, lips, teeth, and so on, modulates the basic excitation signal to produce what we know as a human voice. This can be realized in a similar way using electronics, as shown in Fig. 4-25.

In the electronic realization, the vocal tract is replaced by a complex filter whose transfer function can be rapidly modified. Wiggins and Brantingham used a multi-stage lattice filter with coefficients $K$ to $KN$, which could be changed as and when required to produce the desired filtration characteristic. In this way the synthesis on any given speech word can be realized by specifying, basically, only the following parameters:

1. The type of excitation (voiced or unvoiced)
2. The degree of amplification
3. The filter coefficients ($K$ to $KN$ )
4. The pitch of voiced sounds (essentially determines the impulse repetition rate)

The application of the above technique results in a data rate which is somewhat dependent on the actual word being transmitted. Practical experience has yielded very intelligible speech with data rates lying in the range 600 to 2400 bit/s; somewhat less than the 64 kbit/s required for a classical PCM system!

# 5

# Analog To Digital
# Conversion—Coders

We shall now describe the final step in the production of a PCM signal, that is coding.

The term *coding* in the digital context can unfortunately refer to several processes. For example, we may think in terms of making the message unintelligible to unauthorized users, or alternatively, the conversion of a simple binary signal to another digital form more suitable for transmission (see Chap. 9). Here we shall consider only the simple coding structures and mechanisms that are required to represent a particular analog value digitally.

There are many different types of coding mechanisms, and to simplify our analysis it is useful to classify them. We shall distinguish each coding system as belonging to one of two groups, depending on whether they are based on capacitor-charging or voltage comparison techniques.

The *capacitor-charging* coders are arranged to perform a defined digital process, while a capacitor is under charge. This process stops, in order to reveal the coded word, when the voltage across the capacitor is equivalent to some reference, or the quantum value. The method is usually associated with cheap coder designs, where a small message signal bandwidth is tolerable (see Sec. 5.2).

The *voltage comparison* coders operate by generating discrete voltages, whose levels are equivalent to digital words. Thus it is possible to obtain the coded word directly at the instant when the generated voltage and quantum value are equal. We shall see later that coders belonging to this group can work either in a serial (sequential), or parallel manner, and that some coders are based on a combination of serial and parallel processes. The serial systems sweep the possible amplitude range gradually and must be restricted to uses where only a small message signal bandwidth is required. The parallel systems are expensive to implement but are able to accept wideband signals (see Sec. 5.3).

In the later sections of this chapter we shall consider error correcting codes, and some special code mechanisms that have been developed. First, however, it is necessary to briefly outline the basic principles that make digital coding possible.

## 5.1 BASIC PRINCIPLES

In Chap. 1, Fig. 1-1, we saw how 16 quantum levels could be represented by a sequence of 4 binary pulses. These sequences are of fixed length, and are referred

to as coded *words,* or *characters.*  The example shows how the quantized approximation of an amplitude value may be conveyed as a binary number; in this case of
length 4 binary digits.

The *binary number code* is the best known example of a more extensive class
known as *weighting codes.*  Historically, there have been many other coding classes,
and we shall consider some of these in this chapter; however, since many equipments
today employ weighting codes we shall devote most of our attention to this type.

The binary digit, or bit, defines two logical states denoted as logical 1 or 0.  These
may be very easily specified electrically; for example, a level of 5 volts at the output
to some device may be taken as representing a logical 1, while ground potential
represents a logical 0.  For our purposes these details are unimportant, and in the
descriptions that follow we shall only be concerned with logical representations, not
absolute values.

A few words about *binary counters* are appropriate here, since it is important
that their operation is well understood.  Since most counters are constructed using
individual flip-flops, we shall begin our description with this component.

A *flip-flop* is a device that can maintain equilibrium in either of two states.  We
can represent this mechanically, by a ball at rest in one of two valleys, as shown in
Fig. 5-1.  Here, the ball can only change valleys when subjected to some external
force.  This setup may be implemented electrically by using a bistable transistor
configuration, while the external force may be considered to be a clock pulse.

There are many different types of flip-flop.  The simplest example is the set/reset
device envisaged above; however, more complex schemes based on the master/slave

**Fig. 5-1**  A simple representation of a flip-flop.

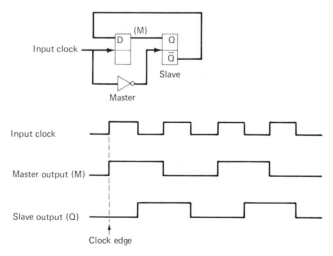

**Fig. 5-2**  A D-type master/slave flip-flop.

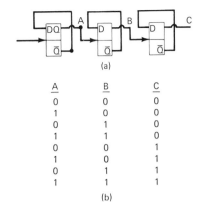

(a)

| A | B | C |
|---|---|---|
| 0 | 0 | 0 |
| 1 | 0 | 0 |
| 0 | 1 | 0 |
| 1 | 1 | 0 |
| 0 | 0 | 1 |
| 1 | 0 | 1 |
| 0 | 1 | 1 |
| 1 | 1 | 1 |

(b)

**Fig. 5-3** (a) A binary counter; (b) the associated truth table.

principle are also commonly used.[1] We consider in Fig. 5-2 the case of a single flip-flop being used as a divide by 2 counter. Each time a positive clock edge occurs, the flip-flop changes state, resulting in half as many pulses (or positive edges) at the $Q$ output compared to the clock input.

The classic binary counter uses several flip-flops cascaded as shown in Fig. 5-3a. The associated truth table for this circuit is defined in Fig. 5-3b, where we have assumed the initial condition at the $Q$ outputs of flip-flops $A$, $B$, and $C$ as logical 000. If now the counter is clocked, then a sequence defined by the truth table will ensue.

When a binary counter is used as part of a coder, it is normal to impose the initializing condition (i.e., 000) just before each sampling instant. Thereafter only the correct number of clock pulses need be supplied to produce the appropriate sequence at $Q_A$, $Q_B$, and $Q_C$, which is used as the basis of the digital code. In the later sections of this chapter we shall describe several methods that produce the correct number of pulses envisaged above.

This description of a binary counter has of necessity been brief. The counters used in coders range from the relatively simple asynchronous ripple types described, to quite complex synchronous designs. The study of these digital circuits is a subject in itself, and is covered by extensive literature.[2] We shall omit these details here.

A coded word of length $b$ binary digits is capable of defining a maximum number of $2^b$ quantum levels. Our earlier analysis (see Chap. 4) showed that for telephony applications a tolerable quantizing error could be obtained, provided a total number of approximately 256 quantum levels, split equally between positive and negative excursions, are specified. This number can be specified using 8 bits ($b = 8$); which implies a binary counter composed of eight flip-flops rather than the three envisaged above.

Some codes employ many more bits than the minimum. These are the error correcting types, which are the subject of Sec. 5.4.1. However, for the moment we shall assume that the chosen coding scheme aims to minimize the number of bits required.

The relationship between consecutive quantum values and the corresponding coded

[1] Texas Instruments IC Applications Staff, *Designing with TTL Integrated Circuits,* McGraw-Hill, New York, 1971.

[2] Texas Instruments IC Applications Staff, *Designing with TTL Integrated Circuits.*

words is varied. Clearly it is possible to assign any combination of the $b$ bits to specify each quantum level, with the only stipulation that no two levels may be represented by identical sequences. In principle then there are $2^b$ possible coding schemes. For example, consecutive levels may conceivably be specified as:

| Level, $b=4$ | Binary code | Code $X$ |
|:---:|:---:|:---:|
| 1 | 0  0  0  0 | 1  0  1  1 |
| 2 | 1  0  0  0 | 0  0  0  0 |
| 3 | 0  1  0  0 | 1  1  1  1 |
| 4 | 1  1  0  0 | 0  1  1  0 |
| 5 | 0  0  1  0 | 0  0  0  1 |

The binary code has a regular predictable format and may be generated easily. On the other hand code $X$ is completely irregular, and is impracticable to generate. Consequently coding schemes must be appraised not only for their own merit, but also for their convenience of generation and detection. For these reasons we shall consider in the later sections of this chapter, first the coding mechanism, and then the coding structure.

The so-called *symmetrical codes* deserve special mention. Many message signals, such as speech, extend more or less equally above and below the quiescent level. In this case it may be advantageous to select the first code bit to denote the sign (positive or negative) of the sample value, while the remaining bits are used to specify the magnitude of the departure from the quiescent level. The resultant code has a symmetrical pattern when viewed about its center, and is easy to generate. There are many examples of such codes, but here we shall confine ourselves to: (1) the symmetrical binary number code and (2) the reflected binary number code (Gray code; see Sec. 5.3.1).

## 5.2   CODERS BASED ON THE PRINCIPLE OF CAPACITOR CHARGING

Coders belonging to this class are of considerable historical interest, since they were employed in several early systems. These equipments were very slow, yet offered a cheap, and inherently simple solution to the coding problem. Over the years the slow speed, and therefore reduced bandwidth, has been unsatisfactory for most applications. Therefore to date, most designs have employed sequential coding techniques (see Sec. 5.3).

In recent years there has been a marked improvement in the speed at which counting operations are feasible. Thus while the early system designers had to be content with counters operating at less than 1 MHz, today it is possible to use frequencies in excess of 500 MHz. For this reason high-speed counting coders, based on the principle of capacitor charging, have lately attracted considerable attention. It is this fact that has earned them a place here.

### 5.2.1   The PPM or PWM Coder

In Sec. 3.1, we briefly outlined a technique that led to the production of a pulse position modulation or a pulse width modulation signal. The described method showed how PPM and PWM signals may be obtained directly by comparing the stored analog values with an internally generated ramp of fixed repetition rate. We shall now extend this treatment, and include the use of PPM signals as an intermediate step within a coder.

Suppose that the slope of the ramp equals $K^{-1}$ (see Fig. 5-4). Then each amplitude sample will be represented using PPM by a pulse displaced in time $t_s$. The displacement is proportional to the analog sample value $a_s$ and is given by

$$t_s = Ka_s \qquad (5.1)$$

The PWM signal is a sequence of pulses of variable width $t_s$. This may be conveniently obtained from the PPM signal by setting a flip-flop at each sampling instant,

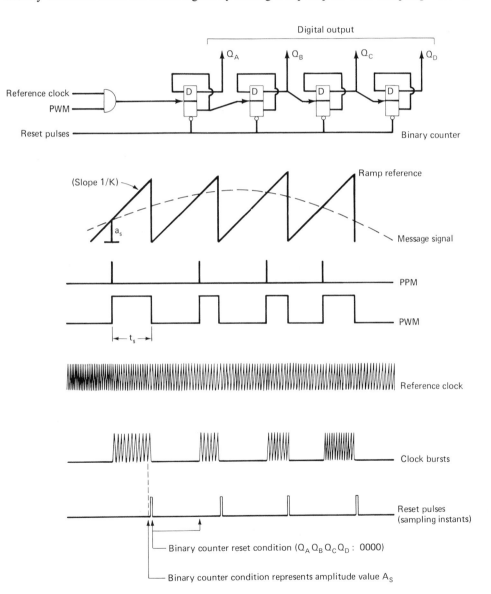

**Fig. 5-4**  The implementation of a coder using PPM or PWM.

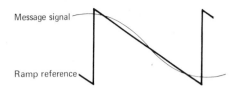

Message signal

Ramp reference

**Fig. 5-5**  Invalid PPM or PWM generation
using a reference ramp.

and resetting it with the PPM pulses.  The output of the flip-flop provides the desired
PWM signal.  The main circuit elements of the coder are shown in Fig. 5-4.

A reference frequency is used to provide a counter clock under the control of
the PWM signal.  Thus, the counter only receives clock pulses during the interval
specified by $t_s$.  In this case the digital count must be proportional to the analog
input value, and represents the required code.  At the end of each count the counter
must be reset to its initial condition, for example: 0000.

It is normally assumed, and indeed is essential to the process, that one and only
one intersection of the ramp and analog message (i.e., stored value) occurs during
the sampling interval $T$.  This implies a frequency limitation over and above that
imposed by the sampling theorem.  The point is illustrated in Fig. 5-5, where although
the sine wave message signal is being sampled at higher than the Nyquist rate, the
ramp intersects three times each period.  In order to avoid this invalid condition,
it is necessary to specify that the slope of the ramp must always exceed that of the
message signal waveform.

Suppose that the message signal is restricted to the range $\pm\frac{1}{2}A$, and that it has
a sinusoidal shape given by

$$y = \frac{1}{2}A \sin \omega t \tag{5.2}$$

Therefore the message signal slope is given by

$$\frac{dy}{dt} = \frac{1}{2}(A\omega \cos \omega t)$$

$$\left(\frac{dy}{dt}\right)_{max} = \frac{1}{2}(A\omega) \tag{5.3}$$

If the peak to peak amplitude of the ramp is taken to be unity, its slope equals
$T^{-1}$.  Consequently, to avoid the invalid condition we must specify:

$$\frac{1}{T} > \frac{1}{2}(A\omega) \tag{5.4}$$

The decoder design is illustrated in Fig. 5-6.  In order to retrieve the message it
is first necessary to generate the PWM signal by synchronizing the decoder and
coder counters.

Suppose that the coder-counter counts from an initial position 0000 to 1011 in
time $t_s$ seconds.  Also that the decoder-counter is set, using the received coded word,
to position 1011, and allowed to count until condition 0000 is reached.  In this
case the decoder-counter functions during the interval $T - t_s$ only and the effective
result is to produce an inverted PWM signal.

The message may be obtained directly from the PWM signal by low-pass filtration, or if lower distortion is required by the insertion of a pulse amplitude modulation (PAM) stage. The conversion of PWM to PAM will be self-explanatory from Fig. 5-6.

It is clear that the ramp supplied at both the coder and decoder must be perfectly complementary, and very well specified, to ensure a distortion free message signal.

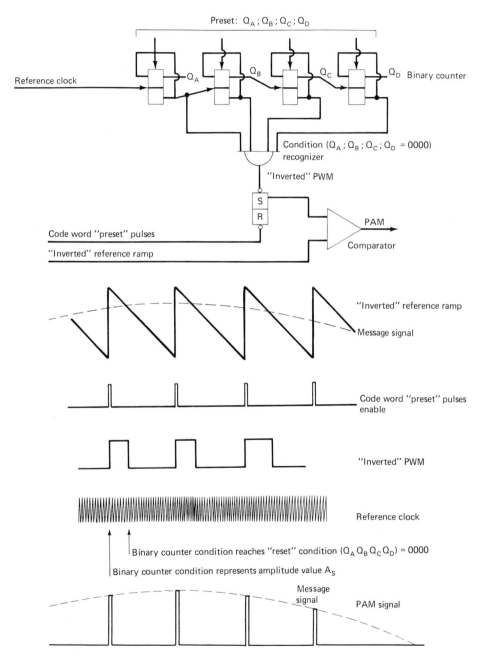

**Fig. 5-6** The implementation of a decoder using PPM or PWM. (The term *inverted* refers to the inverse of signals shown in Fig. 5-4).

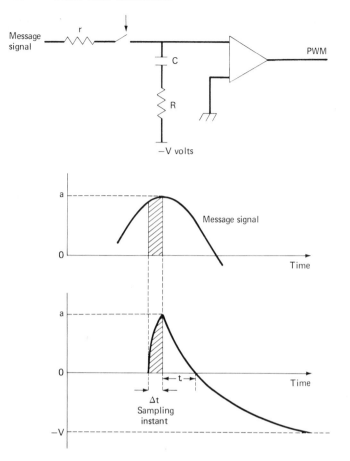

**Fig. 5-7**  The generation of an exponential reference ramp.

An exactly linear sawtooth as envisaged above is a somewhat unnatural process; and in any case because of the advantages offered by a companded system there is some merit in adopting a nonlinear ramp.  Fortunately an exponential decay is one of the most natural waveforms to generate, and it approaches the requirements of an ideal companding law for speech (see Chap. 4).

A suitable circuit for providing an exponential ramp is illustrated in Fig. 5-7, together with the appropriate decay waveforms.  Here an $RC$ circuit is charged to the analog sample value $a$, and allowed to decay towards its asymptote at $-V$ volts. The time $t$ required to discharge the capacitor from $a$ to zero volts is given by

$$e^{-t/RC}(a + V) - V = 0 \qquad (5.5)$$

This may be rewritten as

$$\frac{t}{RC} = \log_e\left(1 + \frac{a}{V}\right) \qquad (5.6)$$

It is convenient to normalize the above expression by choosing a new time variable $t'$, such that $t' = 1$ when $a = 1$.  In this case the time constant $RC$ is given by

$$RC = \left[ \log_e \left( 1 + \frac{1}{V} \right) \right]^{-1} \qquad (5.7)$$

Therefore, Eq. (5.6) may be rewritten as

$$t' = \frac{\log_e (1 + a/V)}{\log_e (1 + 1/V)} \qquad (5.8)$$

This function may be identified with the $\mu$ companding law [Eq. (4.29)] by putting the compression parameter $\mu = V^{-1}$.

The arrangement shown in Fig. 5-7 may only be applied to one polarity of the signal, since it clearly exhibits a different response for positive compared to negative values. When symmetrical signals, such as speech, are involved either two circuits must be provided, or alternatively one may be used within a symmetrical coder. The latter overcomes the problem by coding the sign and magnitude values separately.

Special attention must be taken when defining the charging cycle of the configuration illustrated in Fig. 5-7. The load resistance $r$ must necessarily be very small, since this value defines the charge-up-time constant $rC$ of the circuit. The sampling interval $\Delta t$ must be specified such that the voltage across the capacitance $V_c$ is almost equal to $a$, the sample value, before the decay cycle commences.

### 5.2.2  The Variable Frequency Coder

The basic principle of operation is best illustrated by reference to Fig. 5-8. Essentially, the coder consists of a voltage controlled oscillator (VCO) followed by a binary counter. The stored analog sample value applied to the VCO input dictates the counter clock frequency, and hence the eventual binary count. Clearly for proper operation of this circuit the VCO must have a very rapid response to level changes at its input. The digital count between samples must be proportional to the analog input, and is taken to represent the required code. The counter is reset to its initial state at the start of each cycle.

The decoder circuit performs the reverse operation of the coder, and functions in a similar manner. The physical details are omitted here since they may be easily understood from the description of the PPM/PWM decoder.

### 5.2.3  The Referenced Ramp Coder

This type, although similar in operation to the PWM coder (see Sec. 5.2.1), offers the advantage of increased accuracy, but requires a modest increase in equipment complexity. The block diagram of the coder, together with the appropriate waveforms, is illustrated in Fig. 5-9.

The PWM coder described earlier is inherently inaccurate, due to errors that are introduced during the generation of the ramp. These are caused by

1. The uncertainty of the initial potential $V_u$ across the ramp generation capacitor $C$. This will be dependent on the previous history of the ramp voltage levels.
2. The decay approaching the analog sample value $V_A$. This will be negligible for large values of $V_A$, but not for small ones.

The coder considered here aims to eliminate these errors by performing two integrations rather than just one. The first integration produces a ramp that tends towards $V_A$, and begins at $V_u$, as shown in Fig. 5-9. This integration is allowed to continue for some suitable fixed interval of time $t_F$, whereupon the second integration is selected.

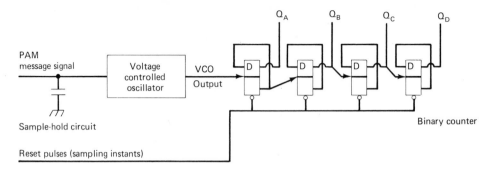

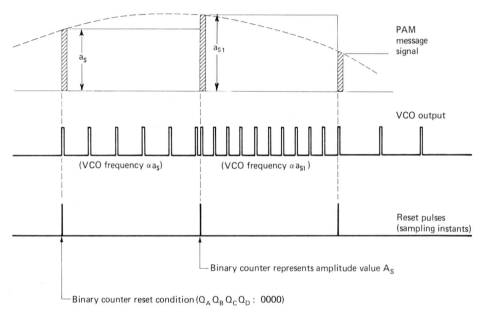

**Fig. 5-8**  The variable frequency coder.

The second reference ramp tends towards a known negative reference voltage $-V_R$, and begins at a voltage defined by the first integration.  The physical selection of the ramps is achieved by switching the integrator input from $V_A$ to $-V_R$, using switches $S1$ and $S2$.

It can be shown mathematically that the time $t_s$ taken for the ramp of the second integration to reach zero volts is proportional to the stored sample value $V_A$.  We might have anticipated this result intuitively, since if $V_A$ is large, then the greater will be the value at the end of the first integration, and the longer it will take for the referenced ramp to return to zero.

A digital counter is used to specify $t_F$ exactly, and to generate a binary code that is proportional to the interval $t_s$.  The timing of events is defined within Fig. 5-9.

The counter is reset to its initial state at the beginning of each cycle, and a suitable reference frequency is used to clock the counter thereafter.  When the count reaches the condition 0000 for example, a time $t_F$ has elapsed and switch $S1$ opens, while

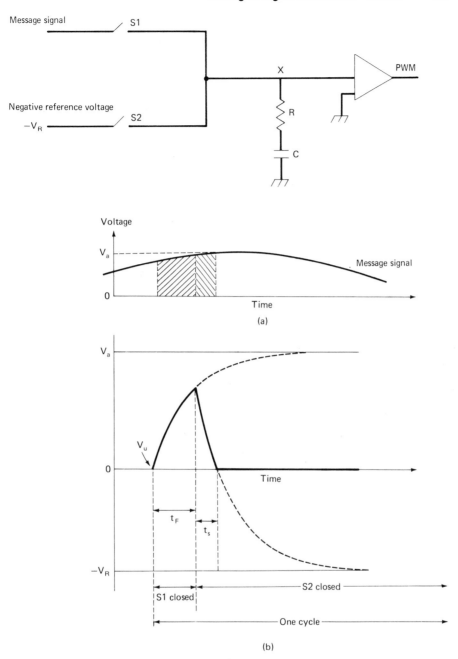

**Fig. 5-9** The referenced ramp coder: (a) the message signal waveform; (b) the charge/discharge waveforms, at point $X$.

$S2$ closes to select the reference voltage $-V_R$. The first integration has now ended and the second begun.

The count is allowed to continue throughout the interval $t_s$, whereupon the referenced ramp crosses the zero-volt threshold and an analog detector inhibits further counting. The value of the final digital count is proportional to the stored analog value $V_A$, and represents the required code.

The integrator output voltage $V_0$ that occurs at the end of the fixed time interval $t_F$ is given by

$$[V_0]_{t_F} = \frac{1}{RC} \int_0^{t_F} V_A \, dt + V_u = V_A \frac{t_F}{RC} + V_c + V_u \tag{5.9}$$

where $V_c = V_u =$ zero, provided the comparator has its threshold at exactly zero volts.

Similarly, the change in voltage during time $t_s$ is given by

$$[V_0]_{t_s} = \frac{1}{RC} \int_0^{t_s} V_R \, dt = V_R \frac{t_s}{RC} \tag{5.10}$$

Since the integration starts and finishes at zero volts:

$$[V_u] = [V_0]_{t_F} = [V_0]_{t_s}$$

Therefore, from Eqs. (5.9) and (5.10),

$$V_A \frac{t_F}{RC} = V_R \frac{t_s}{RC} \tag{5.11}$$

$$t_s = \frac{V_A}{V_R} t_F \tag{5.12}$$

It is interesting to note that $t_s$ is independent of the circuit's time constant $RC$.

The decoder circuit functions in a similar manner to the coder. The physical details are omitted here since they may be easily understood from the description of the PPM/PWM decoder.

## 5.3   CODERS BASED ON THE PRINCIPLE OF DISCRETE VOLTAGE COMPARISON

Certainly the most common principle in use today is that of discrete voltage comparison. We shall see that the implementation may take many different forms, and it is my deliberate aim to identify as many diverse approaches as possible. Currently the fashion is to use parallel coders for situations demanding high-speed operation (e.g., television signals), while the more economical sequential systems are reserved for the low bandwidth messages (e.g., speech signals).

A parallel coder generates the required coded word in a single operation, without any form of scanning process.

A sequential, or serial coder generates each successive digit of the coded word individually. Consequently, the time allocated for the coding of each word must be increased, and the maximum message bandwidth is limited. Furthermore, the choice of coding structure must be restricted to codes that allow any digit to be interpreted without reference to subsequent digits. In practice, however, these drawbacks are unimportant, and sequential mechanisms, due to their low cost, are used extensively.

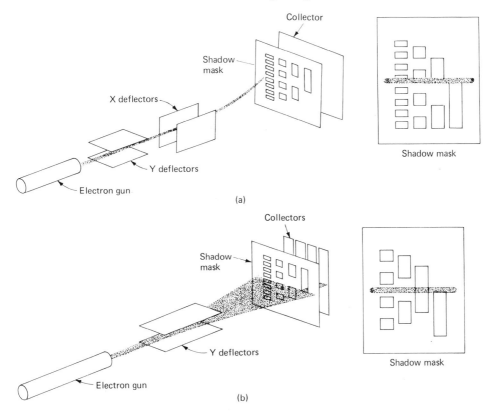

**Fig. 5-10** The electron beam coder: (a) the serial scanning mode; (b) the parallel scanning mode. (From Kenneth William Cattermole, *Principles of Pulse Code Modulation,* Iliffe, London, 1969.)

Perhaps the simplest coding concept to visualize is that based on the electron-beam tube, which was developed and used in the first experimental models at the Bell Telephone Laboratories.[3] A very brief description of this apparatus is given below, since it is not only of significant historical interest but is also helpful to the proper understanding of coding.

### 5.3.1 The Electron-Beam Coder

This coder is similar to a single gun shadow-mask television tube and contains the elements shown in Fig. 5-10. The electron beam emitted by the gun must pass through the holes within the shadow mask before striking one of the collector plates. Now, since striking the collector or not striking the collector is taken to constitute a logical 1 or 0, respectively, the coded word becomes a function of the beam deflection and the holes within the mask. The collectors are arranged such that each specifies a different digit within the coded word.

The vertical deflection of the beam is proportional to the analog sample value, and is caused by the transverse electrostatic field set up by deflection plates $Y$ within the tube.

[3] R. W. Sears, "The Electron-Beam Deflection Tube for Pulse Code Modulation," *Bell System Technical Journal,* vol. 27, 1948, pp. 44–57.

The original coding tubes at the Bell Telephone Laboratories operated in the serial mode (Fig. 5-10*a*). A fine electron beam was deflected along the horizontal axis under the direction of a control signal. This horizontal deflection enabled each digit to be scanned one at a time, until the complete coded word was compiled.

The later coding tubes had a parallel mode of operation (Fig. 5-10*b*). In this case a thin fan-shaped electron beam was adopted. Thus the beam impinges simultaneously on all digit positions, and avoids any requirement for horizontal scanning.

We have already noted in this chapter that a binary number code is ideally suited to coding mechanisms that contain a digital counting stage. The electron beam coder however is better suited to codes that differ by one digit only between levels. Such codes reduce the errors that may occur when the beam is slightly off center with respect to the shadow mask. The problem may be deduced by analysis of Fig. 5-10.

If the beam is slightly misaligned it is possible that the first few digits of the coded word correspond to one quantum level, while the remaining digit of the word relates to an adjacent level. The chosen code structure should aim to keep the error of this misalignment to a minimum.

A common example of a code that exhibits the required property of unit-distance between levels is the *reflected-binary,* or *Gray code,* which is illustrated in Fig. 5-10*b*. This code, devised originally by Elisha Gray for telegraph usage, is generally useful for parallel coders, where errors in determining exact thresholds can easily occur. The Gray code has a naturally occurring symmetrical structure.

The manufacture of coding tubes proved to be difficult and expensive due to the extremely close mechanical tolerances that are required. It is this fact that prevented their usage within commercial equipment. Nevertheless experimental systems have been built to extremely high standards, achieving up to 512 quantum levels, at sampling rates of 12 MHz!

### 5.3.2　Voltage Step-Coder

The voltage step-coder is very simple in concept and design, but has a long response time. It may be used with low bandwidth message signals.

The main elements of the coder are shown in Fig. 5-11. A suitable reference frequency is used to provide a counter clock, while the output to a digital-to-analog (D/A) converter is less than the quantum value with which it is compared. The counter therefore performs two functions; namely: (1) controls the D/A converter, and (2) supplies the required coded word.

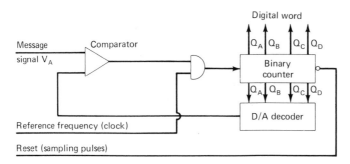

**Fig. 5-11**　The implementation of the voltage step coder.

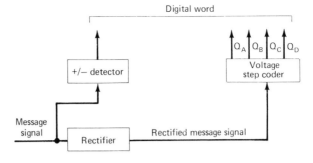

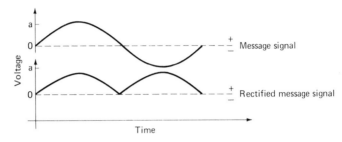

**Fig. 5-12**  The implementation of a symmetrical voltage step coder.

The counter is reset at the start of each sampling interval, whereupon the D/A converter output has its minimum[4] voltage $V_{min}$ level.  The maximum voltage $V_{max}$ occurs after $2^n - 1$ counting steps have been performed where $n$ represents the number of bits in the digital word.  Thus $2^n - 1$ process steps are required before the conversion may be deemed complete, making the technique unsuitable for high-frequency message signals.

D/A converters may assume many different physical implementations.  Typically, various resistive loads may be switched in and out of the circuit, under the direction of the digital code.  Thus the consequent level that is supplied at the D/A output represents the analog equivalent of the coded word.  A more detailed explanation of such circuits is somewhat lengthy, due to the variety of types, and for this reason is omitted here.[4]

We have already noted within this chapter that most message signals are symmetrical about some quiescent voltage $V_Q$ level.  Consequently it is desirable to nominally center the analog scanning range at $V_Q$, and provide either a positive (up) count towards $V_{max}$, or a negative (down) count towards $V_{min}$.  In this case the coder would tend to continuously follow the analog message signal, and would not be reset to $V_{min}$ at the start of each sampling interval.  The resultant code, if this procedure is adopted, is known as the symmetrical binary number code (Sec. 5.1).

A symmetrical step-coder is most easily implemented by interposing a rectifier between the message signal input and the comparator, as shown in Fig. 5-12.  The message signal is first interrogated by a positive-negative value detector, which deter-

[4] David F. Hoeschele, Jr., *Analog to Digital/Digital to Analog Conversion Techniques,* Wiley, New York, 1968.

mines the sign of the sample value; while the rectifier plus voltage step-coder specifies its magnitude.

### 5.3.3  Successive Approximation Coder

The successive approximation coder is also based on the principle of voltage stepping, yet offers a much faster response time compared to the previous method. The code conversion is completed within $n$ process steps.

Each successive digit within the binary code is assigned an equivalent analog weight. In practical terms this means that the first digit of the coded word expresses the quantum level rather coarsely, while subsequent levels add the desired amount of precision.

Suppose, for example, that the coded word consists of 4 digits, and that the start (reset) condition is 0000, equivalent to the minimum voltage $V_{min}$ scanned by the coder. The main functional elements of the coder are identified together with the appropriate waveforms in Fig. 5-13.

The conversion process may be described as follows. The most significant digit is forced to a logical 1, providing the code 1000. The equivalent analog value produced by the D/A converter is then compared to the quantum level $V_A$. If, as in our example, the quantum level $V_A$ exceeds the D/A output, the logical 1 remains at the most significant digit location, and the next most significant digit location is forced to a logical 1 also. Thus, the code becomes 1100.

The described process continues until the D/A generated level exceeds that of the quantum value $V_A$. In our example this occurs when the code becomes 1110. Thereupon the previous logical 1 is removed to produce the code 1100; and then a logical 1 is tried in the least significant digit location, producing the required coded quantum value 1101.

It is important to note that the generated D/A level becomes a better approximation

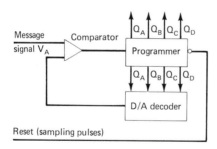

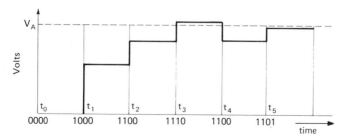

**Fig. 5-13**  The successive approximation coder.

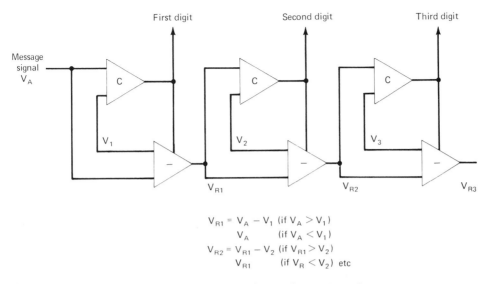

$$V_{R1} = V_A - V_1 \quad (\text{if } V_A > V_1)$$
$$V_A \quad\quad (\text{if } V_A < V_1)$$
$$V_{R2} = V_{R1} - V_2 \quad (\text{if } V_{R1} > V_2)$$
$$V_{R1} \quad\quad (\text{if } V_R < V_2) \text{ etc}$$

**Fig. 5-14** Successive approximation using weighted voltage subtraction.

of the quantum value $V_A$ after each process step. This must be true since the step size is smaller for the least, rather than the most significant digit.

The technique of successive approximation may readily be adapted to segmented companding. The sampled value $V_A$ is in this case first tested using nonlinear analog weighting, rather than the linear scheme previously described. This determines which segment contains the sample value $V_A$. The exact position of the sample $V_A$ within the segment is found by linear analog weighting in the normal way.

The symmetrical version of this coder may be implemented by interposing a rectifier as described in Sec. 5.3.2.

Successive approximation coders have been designed in many different ways other than as just described. An alternative solution based on the principle of successive weighted analog voltage subtraction is worthy of note. The technique is illustrated in Fig. 5-14.

Initially the sample $V_A$ is compared against the analog voltage $V_1$ associated with the first digit of the code word. If the sample $V_A$ is greater than $V_1$, then $V_1$ is subtracted from it and the resultant voltage $V_R = V_A - V_1$ passed to the next comparator. In this case the first digit registers a logical 1.

If, on the other hand, the sample $V_A$ had been smaller than $V_1$, then nothing is subtracted, and $V_A$ is passed as the resultant voltage $V_R'$, to the next comparator. The first digit of the code word is now a logical 0.

The process is repeated at the second, and subsequent comparators, where $V_2$, $V_3$, etc., may, or may not, be subtracted until all the digits of the code word have been compiled.

### 5.3.4 Parallel Coders

Parallel coders are necessarily very complex, and therefore expensive. Since they are extremely fast in operation they are to be found in such applications as in the coding of television signals, or in the conversion of FDM group band messages to the PCM format. A separate comparator must be used for each quantum level specified, in order that a simultaneous conversion can take place. The requirement,

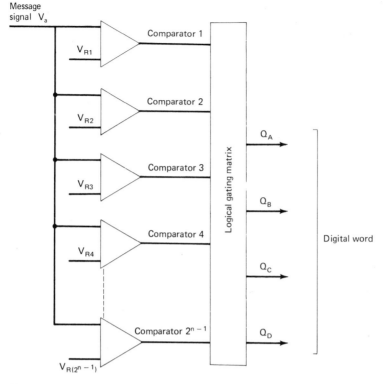

**Fig. 5-15**   The implementation of a parallel coder.

then, is for $2^n - 1$ comparators, and $2^n - 1$ reference voltages, where $n$ refers to the number of digits in the code word.

The comparator output must be reduced to the format of the desired code word. This is achieved by a logic gating matrix as shown in Fig. 5-15.

### 5.3.5  Parallel-Serial Coder

In certain cases neither the high-cost high-speed parallel coder, nor the low-cost low-speed successive approximation coder satisfies the design requirement. A compromise between the two alternatives may offer a satisfactory solution. For example, a segmented companding coder could be implemented in two stages. First, a parallel conversion may be used to define the segment number; this may then be followed by a second stage of coding, using successive approximation to define the exact sample position within a segment.

A reasonably rapid conversion may also be obtained by using a cascaded parallel coder, as illustrated in Fig. 5-16. In this example, two parallel coders are used to specify both the segment number and the position within the segment.

The interesting feature of this design is that the need for several parallel coders within the second stage is obviated by using a programmable attenuator. Thus, once the segment number has been determined, the message signal is amplified or attenuated in fixed steps such that $V_s$ occupies a new, known, reference segment. The position of $V_s$ within the segment remains unchanged however, and the second parallel coder is used to specify this.

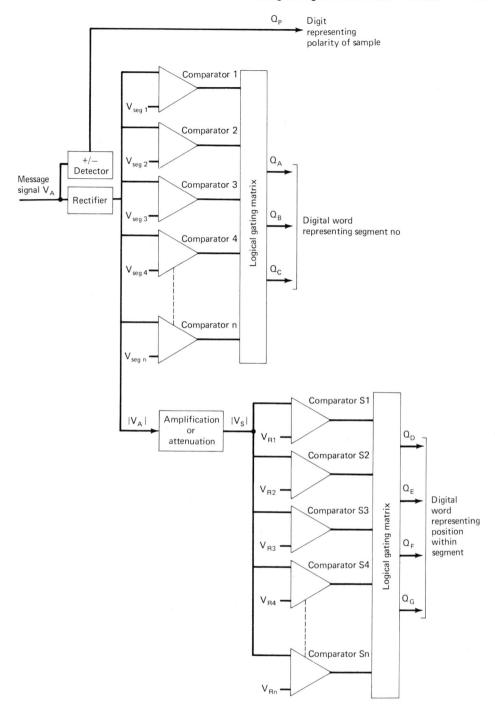

**Fig. 5-16**  The implementation of a parallel serial coder within a segmented companding scheme.

Theoretically, this method could be used to provide very rapid conversions without a large number of comparators. In practice, however, the conversion rate is limited by the slow operation of the programmable attenuator.

## 5.4   DIGITAL PROCESSING

Due to the introduction of cheap readily available digital integrated circuits and microprocessor technology, it has become economically feasible to digitally process information signals prior to transmission and again at the receiver. There are two conflicting reasons for doing this: (1) to detect and correct transmission errors, and (2) to reduce the bandwidth and hence attenuation by removing redundant message information. We shall now briefly consider the implications of this technology.

### 5.4.1   Error Correction

Suppose that we wish to communicate over a transmission link the simple 2-digit words 00, or 11, and that we propose to never allow the words 01, or 10 to be transmitted. In this case, if a single digit error is introduced within the transmission path, the words 00 or 11 could be corrupted to give the disallowed word 01, or 10. Thus, although the receiver realizes that a faulty bit has been introduced, it is impossible to deduce the original message. The adopted code is referred to as *error detecting*.

The same binary information could be communicated by transmitting the 3-digit words 000 or 111, while disallowing the words 001, 010, 100, 110, 101, and 011. It is now possible to perform error correction by ignoring single digit errors. Thus the received word 001 is interpreted as 000, and 101 as 111, etc. In this case a threefold increase of the minimum transmission rate for communicating the binary condition 0 or 1 is required.

If more error protection is desirable the binary message could be transmitted as 00000 or 11111, a fivefold increase in transmission bandwidth. In this case the effect of 2 error digits may be ignored. Thus 11000 is interpreted as 00000, while 11100 becomes 11111.

Codes of this type are commonly used for communicating control messages in cases where faithful transmission is vitally important and the quantity of information (i.e., bandwidth) is small (see Sec. 7.1.2).

The efficiency of an error-correcting code may be considerably improved upon by choosing a more complex coding structure and assuming a truly random noise interference. R. W. Hamming,[5] Bose-Chaudhuri-Hocquenghen, and others have proposed considerably more efficient structures, which are the subject of an extensive literature,[6] and will therefore not be described here. These codes are typically employed when the transmission media is extremely poor, such as in deep outer space, where indeed the noise interference is truly random.

Most terrestrial communications, however, suffer from nonrandom interferences giving rise to so-called *burst errors*. In this case error correction cannot be achieved by the coding structures envisaged above.

The effect of occasional errors within an audio or video signal is minimal since

---

[5] R. W. Hamming, "Error Detecting, and Error Correcting Codes," *Bell System Technical Journal,* vol. 26, 1950, pp. 147–160.

[6] W. W. Peterson and E. J. Weldon, Jr., *Error-Correcting Codes,* 2d rev. ed., M.I.T., Cambridge, Mass., 1972.

the inherent redundancy of such signals is high. The effect of burst errors on data transmission is more serious however, and requires careful study.

Typically, data users are responsible for processing the received signal and removing any erroneous samples. Consequently, data is conventionally transmitted as a finite block of digits followed by a check word, which is dependent on the preceding information block. If the check word indicates a transmission error, then the information block is repeated once more, otherwise it is destroyed and new information sent.

Banks and similar institutions that operate large private data transmission networks commonly correct errors by processing the actual messages using computers; rather than worrying about the transmission mechanism. Thus, for example, the monetary value of a cashed check, the corresponding entry on the customer's statement, and the payment advice slip may be tested for equality. If an error has occurred retransmission of the data is requested.

Such techniques, although cumbersome and time-consuming, are necessary if faithful transmission is to be guaranteed.

### 5.4.2 Bandwidth Reduction

The inherent redundancy within analog message signals suggests that a transmission bandwidth less than so far assumed could satisfactorily be employed. Many techniques have been proposed. However, those involving digital processing seem to offer the greatest possibilities due to the availability of cheap large-scale integrated (LSI) circuit devices.

Coded voice signals have been transmitted at considerably reduced bandwidth by using digital prediction techniques. This method may be described as *predictive quantization* (see Sec. 4.7.3) performed digitally. A major technical drawback is the systems resistance to single digit transmission errors. For example, some experimental equipments require 30 min to resynchronize after each error!

Studio quality color television signals offer more opportunity for bandwidth reduction, compared to voice transmission. A conventionally coded color television signal requires a bandwidth of approximately 120 Mbit/s, while experiments have shown that using digital compression 50 Mbit/s is sufficient. In order to achieve such a reduction the television signal must be stripped into its constituent parts, namely: sound, line synchronization, frame synchronization, blanking, color and video signals. The stripped signals are then recompiled into a form that is more economic in terms of digital transmission rate.

The advantages and disadvantages of the described reduction technique may be summarized as

1. Each system must be tailored to meet the individual properties of the message signal, and is not universally applicable.
2. The reduction in transmission bandwidth must be considered in relation to the increased equipment costs, and may be uneconomic if transmission bandwidth is available cheaply. The advent of cheap special purpose LSI digital processing chips, or on the other hand, the availability of very low cost large-bandwidth transmission media will complicate the economic arguments in the years to come.

# MULTIPLEXING

# 6

# Synchronous Time
# Division Multiplexing

We have seen already that once message signals have been reduced to a digital format, they may be most easily combined, that is multiplexed, with other digital signals by interleaving them in the time domain. The technology involved in this operation is to be the subject of this chapter, and Chap. 7. We shall concern ourselves here, however, only with the basic principles involved while omitting those details concerning asynchronous multiplexing (Chap. 7).

## 6.1  THE INTERLEAVING PROCESS

The fundamental circuit element within any time division multiplexer (TDM) is a *serializer*. This circuit accepts the parallel channel inputs and allows each, taken in turn, access to the output. Thus each channel input is allotted a time slot which uniquely defines the channel information within the serialized output stream. This is illustrated in Fig. 6-1, together with the relevant timing waveform.

We know that the channel messages, in the case of PCM signals, occur as a sequence of fixed-length code words. This gives rise to two different multiplexing procedures. If the channel time slot is long enough to accommodate one complete code word, the multiplexed output signal is termed *word interleaved*. Alternatively each channel code word may be scanned one binary digit (bit) at a time to produce a *bit-interleaved* multiplexed signal. The two processes are illustrated within Fig. 6-2.

The term *character interleaved* is also used, and describes multiplexed telex signals. Each character refers to a code word used to specify the various internationally agreed upon alphanumeric symbols.

The serialization timing is derived from a high stability oscillator, as shown in Fig. 6-1. Typically, the frequency reference used within the oscillator is a quartz crystal of tolerance in the range 10 to 50 ppm (see Chap. 7). This oscillator provides the *master clock* which specifies *all* the timing functions within the multiplexer (see Sec. 6.4).

If all the channel signals are derived from the same master clock, there will be a fixed phase relationship between them, and they are termed *synchronous*. Multiplexing such signals is straightforward. For example, in the case of word interleaving, since the channel word rate always remains equivalent to the information time-slot rate, no allowance for clock tolerances need be made.

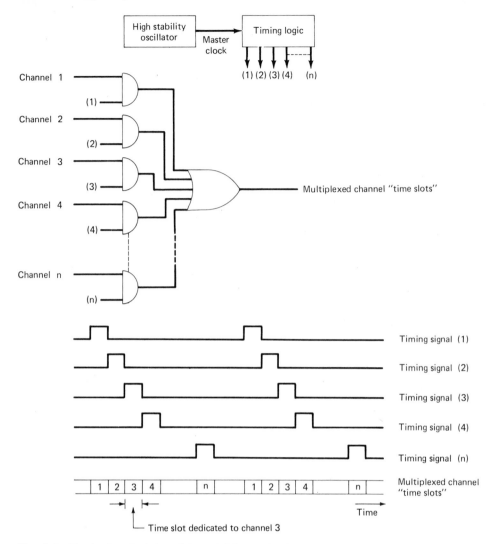

**Fig. 6-1**  The basic principles of time division multiplexing.

Obviously the demultiplexer must be able to identify which time slot is associated with which channel.  For this reason a predetermined recognizable binary sequence is periodically interleaved with the information time slots.  This sequence is used to align, or synchronize the demultiplexer to the multiplexer.  The principle leads to the concept of a *frame,* which is defined as a set of consecutive digit time slots in which the position of each time slot can be identified by reference to a frame alignment signal.

It follows that the multiplexer output signal must possess a bit rate somewhat greater than $c$ times the channel rate if the frame alignment signal is to be accommodated; where $c$ refers to the number of channels.  The most convenient frame structures to implement have a regular format, and for this reason it is desirable to define the *frame alignment word* (FAW) such that it occupies a finite number of time slots exactly. Thus, the number of time slots per frame are typically chosen as $c + 1$, or $c + 2$, etc.

We shall now obtain an expression for the multiplex output bit rate $f_0$ for a frame structure based on $c + 2$ time slots. Let the analog channel messages be sampled at a frequency $= f_s$. Let the number of digits used to represent each coded sample $= w$. Therefore,

$$f_0 = wf_s(c + 2) \tag{6.1}$$

It is important to note that the two time slots allocated for frame alignment may be positioned anywhere within the frame. For example frame formats as specified in Figs. 6-3*a, b,* or *c* are equally valid. However, the frame format should be chosen in such a way as to reduce the average time taken for frame alignment, and to simplify the physical implementation. The term *bunched framing* is used to describe Fig. 6-3*a,* while the term *distributed framing* is reserved for Figs. 6-3*b* and *c.*

Some interleaving schemes do not require additional framing digits, since the tributary signals themselves contain the necessary alignment information. In this case the demultiplexer must scan the incoming digit stream until it detects and synchronizes to some recognizable systematic feature particular to each tributary. Thus the tributary signals define their own channel number and allocated time slot within the digit stream.

Normal pulse code modulated speech does not fulfill the above criterion. However, some television signals converted to PCM, and certain data message switching systems allow the frame to be identified by analysis of the tributary signal.

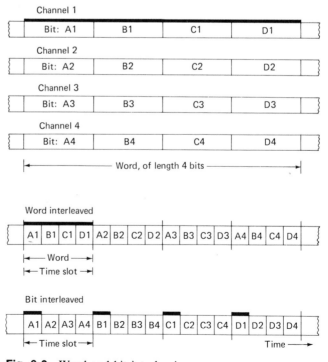

**Fig. 6-2** Word and bit interleaving.

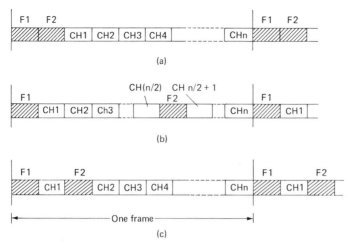

(a)

(b)

(c)

F1, F2: Time slots allocated to unique frame synchronization words, which are exactly repeated each frame

CH1, 2, 3, etc: Time slots allocated to channel message words

**Fig. 6-3**  Frame formats based on bunched (a), and distributed (b, c), frame synchronization words.

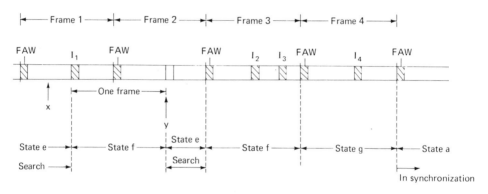

FAW:  Frame alignment word

$I_{1,2,3,4}$:  Information bits that due to their random nature happen to "imitate" the FAW

States a–g:  See Fig. 6-5

**Fig. 6-4**  An example of frame synchronization.

## 6.2  FRAME ALIGNMENT

At this point it is appropriate to investigate the frame alignment process in more detail.  We shall consider the search methods for detecting the proper frame alignment word.  Also, the effect of a frame pattern imitation during the synchronization procedure, and the systems resistance to occasional framing digit errors will be analyzed.

### 6.2.1  The Alignment Procedure

We shall limit ourselves here, initially, to the simplistic frame format illustrated within Fig. 6-4.  In Sec. 6.2.2 we shall show that our analysis may be extended and applied to distributed multiframe alignment schemes with only minor alterations.

Suppose that $F$ digits are used to specify the FAW, and that $N$ digits occupy one complete frame. In this case the number of digits given to information carrying equals $N - F$. We may assume that these information digits exhibit a purely random behavior since we shall impose no control over the channel messages. Thus the probability of a logical 1 at each digit position is 0.5.

Since the information digits may take any form it is likely that the FAW is imitated during the frame by some sequence of $F$ message digits. Furthermore, once the equipment has been aligned, transmission errors can cause either a corruption of an occasional FAW, which may be ignored, or a more severe perturbation that demands the realignment of the system. For these reasons it is necessary to define a state-transition diagram such as shown in Fig. 6-5, and the following alignment states:

State $a$ = Full alignment, system in lock

States $b$, $c$, and $d$ = Provisional alignment, system in check mode

State $e$ = Out of alignment, system in search mode

States $f$ and $g$ = Waiting state, system in search/check mode

There are many different alignment techniques. However, the easiest concept to visualize, and the most commonly used is the step by step, or serial frame alignment procedure. It is instructive to consider an example, using the technique discussed below.

Suppose that the demultiplier is not aligned (state $e$) at a point in time $x$ (Fig. 6-4); the moment when the equipment is switched on. The alignment procedure

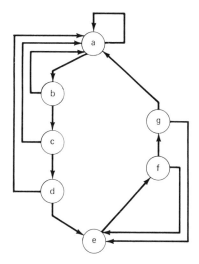

a : In frame synchronization
b : Frame code not detected in frame number n
c : Frame code not detected in frame number n + 1
d : Frame code not detected in frame number n + 2
e : Out of frame synchronism (search mode)
f : Frame code detected in frame number 0 (waiting state)
g : Frame code detected in frame number 1 (waiting state)

**Fig. 6-5** The frame alignment state-transition diagram.

begins by examining the first $F$ digits after point $X$, and comparing them against the FAW. If the digits do not match the FAW, the system slips one digit and interrogates the next sequence of $F$ digits, and so on until a match occurs, whereupon state $f$ is recorded. This process is conventionally referred to as *frame searching*.

In our example (Fig. 6-4), the FAW happens to be imitated by a sequence of information digits at $I_1$. This causes an initial false alignment resulting in state $f$ being registered erroneously.

The presence of the FAW is confirmed each frame. If the test fails the system reverts to states $b$, $c$, $d$, or $e$ depending on the original alignment condition, and the state-transition diagram (see Fig. 6-5). If, on the other hand, the FAW continues to be confirmed each frame, state $a$ will be reached and maintained.

Returning to our example, we note that since the FAW does not occur one frame after the original imitation $y$, the system reverts to state $e$. Thus the frame search cycle recommences at point $y$. The authentic FAW is detected at the start of frame 3, and state $f$ is recorded once more. One frame later the FAW is confirmed (state $g$), and full alignment (state $a$) is registered only when the FAW has been detected for the third consecutive time.

It is interesting, and indeed important to the described process that the imitations $I_2$, and $I_3$ have no effect. Once partial synchronism has been established the system ceases to continue searching for the FAW and merely checks the FAW at one-frame intervals. Systems based on this concept of *serial searching* operate at high bit rates, and are cheap to implement. However, some reduction in the overall mean search time may be obtained by using the more complex solution of *parallel searching*.

It is clear that the risk of simulation of the FAW by random data can be fairly high if $F$ is small in relation to $N$, the number of digits within the frame. In such cases the search time could stretch over many frames before the system settles, and locks to the authentic FAW. Parallel searching reduces the effect of imitations by marking the location of all sequences that match the FAW, and immediately ignoring those that do not repeat at one-frame intervals. After several frames have elapsed all imitations have been ignored, and the position of the FAW is known with certainty. The technique is illustrated within Fig. 6-6.

In this example the FAW is matched at locations 1, 3, 7, 12, 15, and 19 within the first frame. However, since the sequence is not repeated at locations 1 and 9 within the second frame these locations do not contain the FAW, and are consequently ignored.

The parallel alignment procedure rapidly converges to reveal the authentic FAW, while by comparison the serial stratagem continues to lock onto imitations of the FAW throughout several frames. It is interesting to note that there is nothing to be gained between the two systems in the special case when no imitation of the FAW occurs. Furthermore, the advantage in using a parallel stratagem is minimal when imitations of the FAW occur only rarely. Thus the parallel technique is normally reserved for situations when $F$ is small in relation to $N$. This commonly occurs when distributed framing structures are employed.

At first sight it may appear that the problem of imitations can easily be solved by using a very long FAW that is not likely to be simulated by random data. Unfortunately this is undesirable, since the FAW is more easily corrupted by transmission errors (see Sec. 6.2.2).

Tolerance to occasional transmission errors within the FAW is extremely important once the system has achieved full alignment. It should not be necessary to reenact

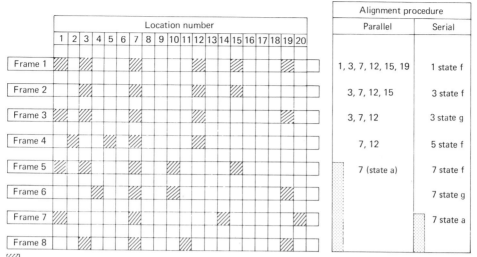

Indicates frame code presence within the given location

**Fig. 6-6**  Frame alignment by parallel searching.

the frame search procedure each time a single digit of the FAW has been corrupted! The states *b, c,* and *d* in Fig. 6-5 provide immunity against these occasional bit errors.  Thus the FAW must be in error during four consecutive frames before a transition to the search mode occurs.  This feature is conventionally referred to as the system's *flywheel.*

Typically transmission errors occur as short bursts rather than being spread out uniformly over a period of time.  These bursts have a duration of a few milliseconds only, and are due to such terrestrial phenomena as lightning strikes, car ignition systems, etc.  The number of steps required to lose alignment, that is the momentum of the flywheel, should be such that the worst error bursts are accommodated.

At this point it is appropriate to consider the described alignment processes quantitatively.[1]  We have already identified the following parameters:

- The number of digits used to specify the FAW $= F$
- The number of digits within one complete frame $= N$

We shall assume a serial alignment procedure, and that frame searching starts at the point $x$, which lies $xN$ digits before the authentic FAW.  The first $F$ digits that occur after $x$ are examined, for which we shall assume a probability of $P(F)$ that the FAW is imitated, and a probability of $S(F)$ that no such simulation takes place. Therefore,

$$S(F) = 1 - P(F) \tag{6.2}$$

During the search $S(F)xN$ tests are made without detecting an imitation.  In these cases the system slips one digit period, and tests the new sequence for the FAW.

Whenever a simulation of the FAW is detected, the system reexamines the digit

[1] H. Haberle, "Frame Synchronization PCM Systems," *ITT Electrical Communications,* vol. 44, 1969, p. 280; and O. Brugia and M. Decina, "Reframing Statistics of PCM Multiplex Transmission," *Electronic Letter,* vol. 5, no. 24, 1969, p. 623.

stream one frame later before continuing the search. Since we may assume that there is no correlation between the test and check cycle in the case of imitations, then $P(F) \times N$ tests need to be made before the authentic FAW is detected.

The time taken to discard each imitation $T$ equals the time occupied by one frame; provided, of course, a second imitation does not occur exactly one frame after the first. The probability of a second imitation not occurring is given by $S(F)$, thus the time spent locking to nonrepeated imitations $T_I$ is given by

$$T_I = \frac{P(F) \times N}{S(F)} \frac{1}{\text{frame rate}} \tag{6.3}$$

We shall assume that multiple imitations spaced at one-frame intervals do not occur. In this case the total frame search time $T_F$ is given by

$$T_F = T_I + \text{time spent slipping } xN \text{ digits}$$

$$T_F = \frac{1}{\text{frame rate}} \left( \frac{P(F) \times N}{S(F)} + x \right) \tag{6.4}$$

If the synchronization word of length $F$ digits is detected in purely random data, then the probability of a logical 1 or a logical 0 occurring at each time slot is 0.5. Thus the probability of a sequence of $F$ digits simulating the FAW is given by

$$P(F) = (0.5)^F = \frac{1}{2^F} \tag{6.5}$$

Substituting Eqs. (6.2) and (6.5) into Eq. (6.4) and rearranging terms gives:

$$T_F = \frac{x}{\text{frame rate}} \left( \frac{N}{2^F - 1} + 1 \right) \tag{6.6}$$

This is an important result and is used extensively in determining the adequacy of proposed frame formats, from the alignment point of view. The curve defined by Eq. (6.6) is plotted in Fig. 6-7, where $x$ has been assumed equal to one; and the FAW length $F$ is expressed in terms of $\alpha$, the fraction of the total transmitted information used for synchronization. Therefore,

$$\alpha = \frac{F}{N} \tag{6.7}$$

The plotted curves exist between two boundaries. The upper boundary defines the limiting case where only one digit per frame is given to synchronization $F = 1$; alignment cannot be achieved for values of $F$ less than one. The lower boundary indicates that the shortest possible synchronization time must be at least equal to the time occupied by one frame. A shorter time is not possible since we have assumed $x$ equal to unity.

Analysis of Fig. 6-7 leads us to the following generally valid conclusions:

1. The alignment time decreases as the percentage of digits given to synchronization increases (frame length = constant).
2. The alignment time is dependent on the frame length if the percentage of digits given to synchronization remains fixed. This is particularly true for small values of $\alpha$.

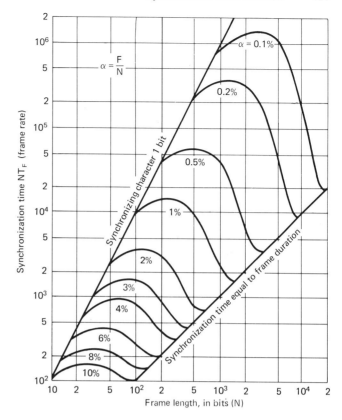

**Fig. 6-7**  Approximate mean synchronization time as function of the frame length. (From H. Haberle, "Frame Synchronization PCM Systems," *ITT Electrical Communications*, vol. **44**, no. 4, 1969.)

3. The alignment time has a minimum value corresponding to some particular value of frame length ($\alpha$ = constant).

Suppose that a frame format is required for the transmission of PCM coded voice signals.  Normally, subjective testing must first be carried out to determine a suitable value for the frame alignment time, say 2 ms.  The smallest value of $\alpha$ that satisfies the required alignment time may then be read directly from one of the curves shown in Fig. 6-7.  The minimum value of the selected curve defines the optimum frame length, while the optimum length of the FAW is obtained using Eq. (6.7).

## 6.2.2  The Frame Alignment Word

The choice of the FAW can have a considerable effect on the overall system performance.  This can easily be understood by considering the operation of the FAW detection circuitry.  Typically a shift register is used as illustrated in Fig. 6-8.

In the diagram a FAW of length 7 digits is allowed to traverse the register 1 bit at a time.  The FAW should be chosen such that the output to the detection circuitry changes only when the word completely fills the register, irrespective of the digits that precede or follow it.

Thus, the requirement is for a FAW that cannot be imitated by itself displaced in time. For example, some suitable words are 1110000, 1111000, and 1110100, or their inverse; while words of the type 1111111 or 0111111 should be avoided. It is instructive to note that if the adopted FAW had been 1111111, for example, an imitation could have occurred at any of the detection steps 1 to 7 (see Fig. 6-8).

Let us consider the limiting case when the system is out of alignment by an amount equal to one time slot, and that the chosen FAW is 1111111. In this case the probability of an imitation of the FAW occurring is dependent only on a single random digit, as shown in Fig. 6-9, and equals 0.5 rather than $(0.5)^F$ as was previously derived. Clearly, similar arguments can be made for other degrees of overlap, and the conclusion

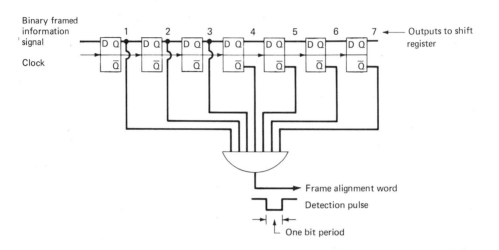

| | Bits outside shift register | | | | Shift register outputs | | | | | | | | Recognition status |
|---|---|---|---|---|---|---|---|---|---|---|---|---|---|
| | | | | | 1 | 2 | 3 | 4 | 5 | 6 | 7 | — | |
| | | | | | Pattern enters shift register | | | | | | | | |
| 0 step | 0 | 0 | 0 | 0 | | | | | | | | | |
| 0 step | 0 | 0 | 0 | 0 | * | * | * | * | * | * | * | * | Pattern not detected |
| 1st step | 1 | 0 | 0 | 0 | 0 | * | * | * | * | * | * | * | Pattern not detected |
| 2d step | 1 | 1 | 0 | 0 | 0 | 0 | * | * | * | * | * | * | Pattern not detected |
| 3d step | 1 | 1 | 1 | 0 | 0 | 0 | 0 | * | * | * | * | * | Pattern not detected |
| 4th step | * | 1 | 1 | 1 | 0 | 0 | 0 | 0 | * | * | * | * | Pattern not detected |
| 5th step | * | * | 1 | 1 | 1 | 0 | 0 | 0 | 0 | * | * | * | Pattern not detected |
| 6th step | * | * | * | 1 | 1 | 1 | 0 | 0 | 0 | 0 | * | * | Pattern not detected |
| 7th step | * | * | * | * | 1 | 1 | 1 | 0 | 0 | 0 | 0 | * | Pattern detected |
| 8th step | * | * | * | * | * | 1 | 1 | 1 | 0 | 0 | 0 | 0 | Pattern not detected |

\* – Represents a logical 1 or 0 (random information assumed)
1 – Represents a logical 1
0 – Represents a logical 0

**Fig. 6-8**  Frame alignment word recognition, using a shift register.

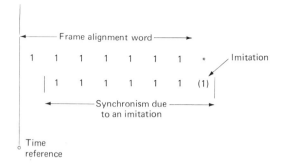

**Fig. 6-9**    False synchronism, due to a poor selection of the FAW.

is that an inappropriate choice of the FAW will result in a reduced performance in the overlap region.

A pattern such as 1110000 cannot be matched by random data, when it is displaced in time. Consequently this pattern has an improved performance in the overlap region, compared to random data.

Previously we considered that the FAW could be matched at any of the $N$ tests made within the frame. Clearly this is not true for a properly chosen FAW since only one match is possible within the overlap region. Thus the number of tests per frame is effectively reduced by an amount $F-1$ equivalent to the length of the overlap. In this case Eq. (6.4) may be rewritten as

$$T_F = \frac{x}{\text{frame rate}} \left\{ \frac{P(F)[N-(F-1)]}{S(F)} + 1 \right\} \tag{6.8}$$

Substituting Eqs. (6.2) and (6.5) into Eq. (6.8) and rearranging terms gives:

$$T_F = \frac{x}{\text{frame rate}} \left( \frac{N+1-F}{2^F - 1} + 1 \right) \tag{6.9}$$

Several studies have been carried out to determine the optimum choice of the frame alignment sequence. The desired patterns must resist imitation by a displacement in time, and in addition must be resistant to occasional digit errors. For these reasons the patterns' performance may be conveniently assessed on the basis of their autocorrelation functions.

### 6.2.3  Multiframe Alignment

Sometimes the frame length demanded by the message signals themselves may be extraordinarily long. An example of such a situation is illustrated in Fig. 6-10$a$, where in addition to the normal channel information, additional data messages must also be transmitted. Typically this extra data may contain details describing the tributary performance, or alternatively low-frequency signaling (dialing) signals associated with each channel, and so on.

The simplistic frame structure proposed in Fig. 6-10$a$ demands a very long FAW if the proportion of digits given to synchronization $\alpha$ is to be maintained. This should be avoided for the following reasons:

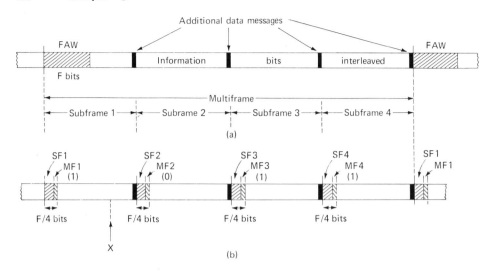

SF1, SF2, SF3, SF4:  Repeated subframe alignment word
            MF1:  Unique part of multiframe alignment word
     MF2, 3, 4:  Unique part of multiframe alignment word

**Fig. 6-10**  Multiframe alignment format.

1. A very long FAW is most easily corrupted by occasional transmission errors. This leads to the FAW being improperly recognized either during the initial search, or during the check cycle, and causes an extended realignment time.
2. If an imitation occurs despite the FAW being very long, the alignment time is likely to be unacceptable. Moreover, individual alignment times could be extremely long.
3. A very long FAW requires a considerable increase in the size of the tributary store.

The total storage required per tributary may be assessed as:

$$\text{Storage, per channel} = \frac{\text{number of consecutive control digits}}{\text{number of channels}} \qquad (6.10)$$

A general term *consecutive control digits* is used here to indicate those digits not assigned to carrying tributary information. The function of the store, then, is to act as a buffer between the continuous stream of tributary information and the corresponding tributary time slots.

A reduction in the channel storage and an improvement in the other shortcomings associated with bunched frame alignment patterns may be obtained by adopting a distributed frame alignment scheme. This leads to the concept of multiframe and short-frame synchronization stratagems.

Suppose that previously a bunched FAW comprising 32 consecutive digits was employed. We may split the original long frame into four separate subframes, each identified by an identical sequence of 7 consecutive digits, while a further 4 digits are reserved to specify the complete multiframe. In this case the proportion of digits given to synchronization $\alpha$ remains the same as before. The new frame structure is illustrated in Fig. 6-10*b*.

Let us assume that the system is out of alignment once more at a point in time

$X$ (Fig. 6-10$b$). Initially, it is necessary to achieve short-frame alignment by matching the 7-digit subframe alignment word (SF), and checking it at subframe intervals. Thus the general principles outlined for bunched framing structures apply equally well to this subframe alignment scheme.

Multiframe alignment may be found by testing the single digits at $MF$ 1, 2, 3, and 4, which uniquely identify the individual short-frames.

The total frame alignment time $T_{MF}$ is given by the sum of the time taken to achieve short-frame alignment $t_{SF}$ and multiframe alignment $t_{MF}$, which may be regarded as separate processes. From Eq. (6.6) we get:

$$t_{SF} = \frac{x}{\text{subframe rate}} \left( \frac{N}{n} \frac{1}{2^f - 1} + 1 \right) \tag{6.11}$$

where $n$ = number of subframes in the multiframe
$N$ = number of digits in the multiframe
$f$ = number of digits within the subframe alignment word $SF$

The search for the multiframe alignment word may be considered free of imitations, provided the alignment pattern is chosen in accordance with the rules set out in Sec. 6.2.2. Therefore,

$$t_{MF} = \frac{1}{\text{multiframe rate}} = \frac{n}{\text{subframe rate}} \tag{6.12}$$

From Eqs. (6.11) and (6.12) we obtain the result that

$$T_{MF} = \frac{1}{\text{subframe rate}} \left( \frac{N}{n} \frac{x}{2^f - 1} + x + n \right) \tag{6.13}$$

### 6.2.4 Distributed Frame Alignment

Some alignment schemes distribute the frame alignment pattern as single digits at regularly spaced intervals throughout the frame. This is an extension of the multiframe alignment concept.

Let us suppose that a frame alignment word consisting of $F$ digits is distributed as single digits uniformly spaced at intervals located $z$ digits apart. The frame search mechanism then tests $z$ sets of digits until the particular set containing the alignment information is located.

On average it is necessary to test $(z - 1)/2$ sets before the alignment word is captured. Each set contains $N/z$ digits, where $N$ represents the total number of digits within the frame. Using similar arguments to those established earlier[2] the result is obtained that

$$\text{Time to locate FAW set} = \frac{x}{\text{frame rate}} \left( \frac{N}{z} \frac{P(F)}{S(F)} + 1 \right) \frac{z-1}{2} \tag{6.14}$$

Once the correct set has been located, a further $xN$ digits must be tested before the frame alignment procedure is complete. Thus the total alignment time $T_D$ in the case of a distributed system is given by

$$T_D = \frac{x}{\text{frame rate}} \left( \frac{N}{z} \frac{P(F)}{S(F)} + 1 \right) \left( \frac{z}{2} + \frac{1}{2} \right) \tag{6.15}$$

[2] P. Bylanski and D. G. W. Ingram, "Digital Transmission Systems," *Institution of Electrical Engineers Telecommunication Series 4*, Peregrinus, Stevenage, England, 1976.

## 6.2.5   The State-Transition Diagram

We have already introduced the concept of a state-transition diagram within Fig. 6-5, and the associated description given in Sec. 6.2.1.   Here we shall attempt to consider the technique of FAW checking in more detail.

Returning to Fig. 6-5 we may note that the transition probabilities are interdependent.   For example, we may decide to employ a long FAW in order to maximize the probability of the transfers $e$ to $f$, $f$ to $g$, and $g$ to $a$, while minimizing the probability of imitations corresponding to the transfers $f$ to $e$ and $g$ to $e$.   Unfortunately, however, a long FAW is more liable to be corrupted by transmission errors; consequently the probability of the transfers $f$ to $e$ and $g$ to $e$ also increases.

An alternative solution is to use a short FAW, and to ignore occasional corruptions that occur once the system is deemed to be in lock.   In this case, frame searching will recommence only if corruptions happen repeatedly.   This technique is commonly used for detecting misalignment, but normally takes too long for use during the initial alignment phase.

For these reasons the chosen state-transition diagram must be a compromise between the parameters identified above.   For a detailed analysis of various schemes, the reader is referred to Bylanski and Ingram.[3]

## 6.2.6   Frame Alignment within a Multiplex Hierarchy

Frequently multiplexers are required to operate in conjunction with other multiplexers and line transmission equipment.   We may consider, for example, the hierarchy defined within Fig. 1-4a.   In this case disturbances such as a loss of synchronism occurring at a higher level within the hierarchy will tend to propagate down to the lower levels.   Thus the total frame alignment time of a primary multiplex becomes the sum of the individual times taken to synchronize the various equipment within the hierarchy on which it is dependent.   It is important to note that the time taken to trigger timing extraction circuits (dependent repeaters), and the time taken for phase-locked loops (asynchronous multiplexers) to achieve lock, must also be allowed for.   Consequently it is usual to demand a more rapid realignment of higher-order equipments compared to those of lower order.

## 6.3   HOUSEKEEPING BITS

Multiplexers designed for telephony applications are required to convey and respond to several network supervision signals.   This facility is permitted by designating certain spare, or unused bits, for communicating the supervision information.   These are called *housekeeping bits*.

Most communication problems are bidirectional, as shown in Fig. 6-11.   In this case, it is desirable to provide a supervision system that indicates to each demultiplexer that the multiplexed signal in the reverse direction was properly received at the far terminal.   Thus if a fault occurs at point $c$, the demultiplexer at site $B$ will lose synchronism, and indicate an alarm (see Sec. 6.4).

The nonoperation of the demultiplexer $B$ should be communicated to site $A$ via a designated housekeeping bit allocated by the multiplexer $B$.

Time division multiplexers used for combining telex or data signals frequently convey considerably more housekeeping information than just loss of synchronism.

---

[3] Bylanski and Ingram, "Digital Transmission Systems."

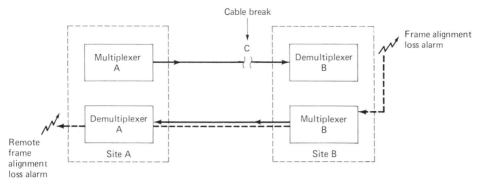

**Fig. 6-11**   Bidirectional communication and system alarms.

The telex (or data) terminal may transmit the following information to the multiplexer:

1. Loss of synchronism at the telex or data machine
2. Loss of power supply feeding (wetting current) at the telex or data machine
3. Character word length and tributary bit rate selected
4. Terminal shutdown; nonoperative for maintenance purposes, etc.

These are communicated to the demultiplexer site using designated housekeeping bits. We shall see later that TDM's designed for telex purposes exclusively tend to be design limited by economic considerations. For this reason a purely telex TDM will normally communicate items (1) and (2) only.

## 6.4  IMPLEMENTATION

The techniques of sampling, companding, coding, serialization, and frame synchronization may now be brought together to form our specified aim: a time division multiplex for speech signals. The block schematic of this unit is shown in Fig. 6-12. For convenience the circuit blocks are arranged into groups designated as channel, or common-equipment circuitry, in addition to the classification of the multiplexing and demultiplexing parts. This form of partitioning is conventional practice, and is based on the following empirical rules:

1. Designated channel circuitry should never be placed in conjunction with that of the common equipment. Thus faulty operation at the channel level will not interfere with the overall functioning of the equipment.
2. Designated channel circuit boards should be identical for all channels. Thus the number of different circuit boards within the TDM may be kept to a minimum.
3. The alignment, initial commissioning, and fault identification procedures should be simplified by the chosen system partitioning.
4. As far as is possible, complete functions should be performed within a single piece of circuitry, and not shared between several boards. This limits the number of interconnections between boards.
5. The mechanics, circuit board size, and permitted power dissipation will influence the choice of the system partitioning.

Several features of the TDM illustrated in Fig. 6-12 are worthy of further comment. We have already in this chapter referred to the need for an accurate and stable

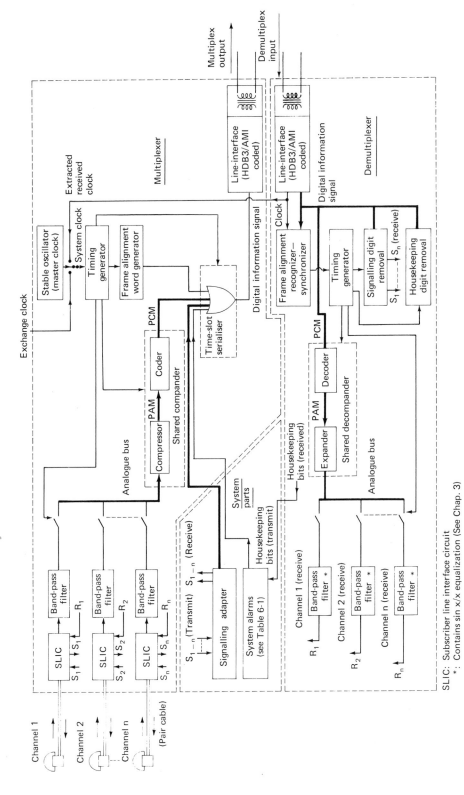

**Fig. 6-12**  Block diagram of a time division multiplex based on a "shared" codec.

SLIC: Subscriber line interface circuit
 *: Contains sin x/x equalization (See Chap. 3)

**Table 6-1   CCITT Recommended Alarms**

| Alarm type | Condition indicated |
|---|---|
| Outgoing signal loss (OGSL) | Indicates the absence of a signal at the multiplexer output |
| Incoming signal loss (ICSL) | Indicates the absence of a signal at the demultiplexer input |
| Frame alignment loss (FAL) | Indicates the nondetection of the frame alignment word |
| Remote frame alignment loss (RFAL) | Indicates the FAL has occurred at the far terminal |
| High error rate (HER) | Indicates that the transmission path is causing an unacceptable level of errors; typically this may be 1 error in $10^3$ or $10^4$ binary digits |
| Encoder fault (ENCOD) | Indicates the malfunction of the encoder |
| Alarm inhibit signal (AIS) | Indicates that a special characteristic signal has been received and inhibits the local alarms |

*system clock.* Normally an internal master oscillator satisfies this requirement; however, if the TDM must operate synchronously with other equipments an *external network clock* should be used. By this means all the multiplexers, and any associated equipments will operate in harmony; that is synchronously with the network clock. In addition, the facility of using the *received timing signal* as the multiplexer system clock should be provided. This feature permits the synchronization of equipments at the remote terminals of a synchronous network.[4] Finally, for use when synchronizing adjacent equipments, the system clock should be separately available at one of the multiplexer outputs.

A simple line interface circuit conventionally converts the multiplexer output, and the input to the demultiplexer, into a ternary signal. This permits interconnection with other equipments over short distances [say 200 meters (m)] using a single cable. Furthermore, by using transformers at the output and input stages greater system flexibility is offered, since in this case the interconnected equipment may use different earth points. Typically these interfaces employ line codes such as high-density bipolar 3 (HDB3), alternate mark inversion (AMI), or scrambled AMI, which are described within Chap. 9.

The designers of commercially produced TDMs must prepare their units so as to facilitate the rapid location of faulty circuits. For this reason, readily accessible breakpoints are incorporated within the line interface. Thus, when a fault occurs the multiplexer may be looped back to the local demultiplexer and disconnected from the network during maintenance periods.

The provision of *alarms* naturally promotes the rapid diagnosis of faults occurring within a telephone network. Consequently the CCITT recommends that the alarms listed in Table 6-1 be included within all multiplex equipments. The implementation of these alarms may be accomplished as discussed below.

The outgoing signal loss (OGSL) and incoming signal loss (ICSL) alarms may be represented by the logical output derived from retriggerable monostables whose inputs are the monitored signal points. Thus provided the monostable time constant is suitably long the continued absence of a signal will signify the alarm condition.

The frame alignment loss (FAL) and remote frame alignment loss (RFAL) alarms may be equated with the synchronization states *e, f,* and *g* identified in Fig. 6-5.

A high error rate (HER) alarm may be derived in several ways, depending on the type of equipment involved. The conventional practice adopted in the case of

---

[4] Stiffler, *Theory of Synchronous Communications,* Prentice-Hall, Englewood Cliffs, N.J.

TDMs[5] is to interpret the rate at which the frame alignment word is corrupted in terms of the overall transmission error rate. The *corruption rate* may be obtained directly by recording the frequency of occurrence for the synchronization states *a, b, c,* and *d* (see Fig. 6-5). The overall transmission error rate can then be calculated using some suitable multiplication factor as explained below.

Let the probability of a single bit being in error within a sequence of $F$ bits = $P_F$, the probability of any bit being in error = $P_B$, and the number of bits within the frame code = $F$. Therefore, the probability of any bit not being in error = $1 - P_B$.

The probability of no error occurring within the $F$ bit sequence, assuming that there is no correlation between the occurrence of errors, = $1 - P_B$. Thus

$$P_F = 1 - (1 - P_B)^F \tag{6.16}$$

If, as is (hopefully!) the case, $P_B$ is very small, Eq. (6.16) approximates to

$$P_F \simeq 1 - (1 - FP_B)$$
$$P_F \simeq FP_B \tag{6.17}$$

Consequently, the required multiplication factor is equivalent to the number of digits used within the frame code.

Subjective tests have shown that a randomly occurring error rate of 1 in $10^5$ bits is the threshold at which a degradation in the quality of speech is not noticeable. An error rate of 1 in $10^4$ bits continues to provide satisfactory quality, while error rates higher than 1 in $10^3$ are unacceptable. Consequently, the HER alarm may be used as a form of early warning system indicating a gradual worsening of the transmission quality (see Chap. 8).

The encoder fault (ENCOD) alarm aims to indicate the malfunction of either the coding or the decoding circuitry by proving their operation when the encoder would otherwise be idle. This occurs during those time slots allocated to the transmission of the frame alignment word, or control signals. During this time interval the decoder output is connected to the coder input, as shown in Fig. 6-13.

The test is performed by injecting into the decoder a preselected code word which is equivalent to the voltage $V_T$. After expansion the voltage $V_T$ becomes $V_{TE}$, as shown in Fig. 6-13. The compressor, and coder, if all is functioning well, performs the reverse operation resulting in the original code word being reproduced at the coder output. Thus, the proper operation of the complete companding encoder may be verified by comparison of the injected and retrieved code words.

The test word must be selected with care, and take account of the practical limitations set by the coding circuitry. A small DC offset is likely to be introduced during the encoding process. Consequently the least significant digits of the retrieved code word may be in error, and are ignored for the purpose of comparison.

The effect of introducing a DC offset is illustrated within Fig. 6-14. In the example the injected test word $A$ is corrupted to a new word $B$, and similarly word $C$ degenerates to $D$, for the same degree of offset. However since $C$ lies adjacent to a segment boundary, the offset corrupts both the least significant digits and the segment digits (word $D$), rendering the comparison ($C$ to $D$) impossible. Thus test words that correspond to a point at the center of the segment must be employed.

The alarm inhibit signal (AIS) indication is used to show that a particular equipment

---

[5] For transmission links, violations in the transmission coding scheme are typically used (see Chap. 8).

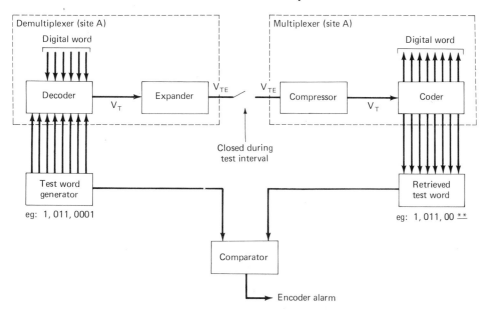

Note least significant digits $\underline{**}$ are ignored in the comparison

**Fig. 6-13**   Testing the encoding and decoding operations.

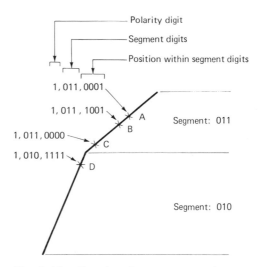

**Fig. 6-14**   Choosing the most appropriate test pattern for testing the coder and decoder.

is out of service due to a fault elsewhere in the network.  This feature causes maintenance engineers to be summoned only when required, and allows them to ignore
faults that occur within equipments outside their area of responsibility.  It is instructive
to analyze the effect of the AIS within a multiplex hierarchy.

Suppose that a higher-order multiplexer detects a common equipment failure.
In this case the AIS is transmitted at all tributary outputs, and is eventually detected

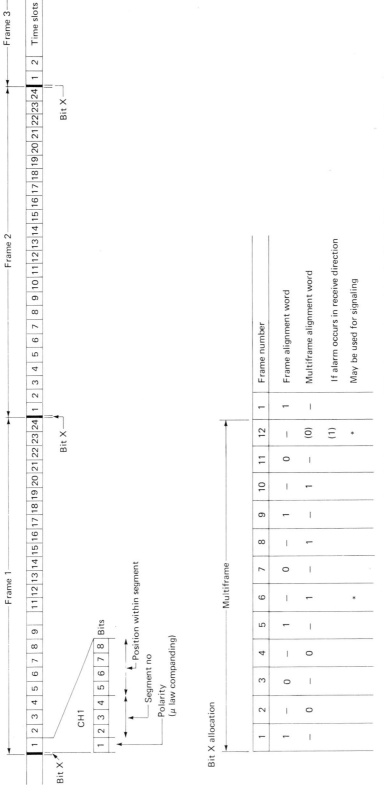

**Fig. 6-15** The frame format of the Bell T1, 24-channel multiplex system. (This format is used in the United States, Canada, and Japan.)

at the dependent multiplex equipments.  This causes an alarm indication at the higher-order multiplex only; the point where the fault originally occurred.

## 6.5  PRACTICAL SYSTEMS

In conclusion to this chapter it is appropriate to briefly review the most important synchronous time division multiplexers in common usage today.  The following examples have been selected.

1. The Bell T1 system: 24 voice channels (CCITT recommendation G733).
2. The European Primary system: 30 voice channels (CCITT recommendation G732).

### 6.5.1  The Bell T1 TDM

The frame format adopted for this North American system is shown in Fig. 6-15.  Each of the 24 voice channels are sampled at 8 kHz, compressed using the $\mu$ law and coded as an 8-digit word.  The resultant channel messages are word interleaved to form an uninterrupted sequence of 192 digits.

A single framing digit (bit $X$) is inserted at the start of each sequence, giving a total frame length of 193 digits.  The multiframe comprises 12 frames, as shown in Fig. 6-15.

Both the frame alignment word and the multiframe alignment word (MFAW) are distributed throughout the multiframe within the bit $X$ digit positions.  In this case it is necessary to examine 6 digits, that are separated by 385 digits from each other, in order to match the FAW.  This makes the implementation of the frame alignment circuitry somewhat complex, since parallel searching techniques must be employed unless extraordinarily long search times are permissible.  Once the FAW has been matched, only a small amount of additional circuitry is required to detect and align the equipment to the MFAW.  The mean search time for this system is of order 24 ms.

The bit $X$ position within frame 12 is normally fixed at a logical 0.  However, if the associated demultiplexer loses frame alignment this bit changes to a logical 1, causing a RFAL alarm to be transmitted to the far terminal (see Sec. 6.4).

A comprehensive description of this equipment, including details relating to the signaling system which have been omitted here, is given by C. G. Davis.[6]

### 6.5.2  The European Primary TDM

This equipment is commonly referred to as a *primary, 30-channel,* or *32 time-slot* multiplexer, for reasons that are obvious.  The frame format is shown in Fig. 6-16.

The TDM combines 30 voice channels that have been sampled at 8 kHz, compressed using the *A* law, and coded as an 8-digit word such that:

| | |
|---|---|
| Digit 2 | Indicates a positive (logical 1), or negative (logical 0) amplitude value |
| Digits 2, 3, and 4 | Indicates the segment occupied by the sample |

---

[6] C. G. Davis, "An Experimental Pulse Code Modulation System for Short Haul Trunks," *Bell System Technical Journal,* vol. 41, 1962, pp. 1–24; and International Telegraph and Telephone Consultative Committee, *CCITT Orange Book,* vol. III-2, G711 (recommendation), G733 (United States system), International Telecommunications Union, Geneva, Switzerland, 1977.

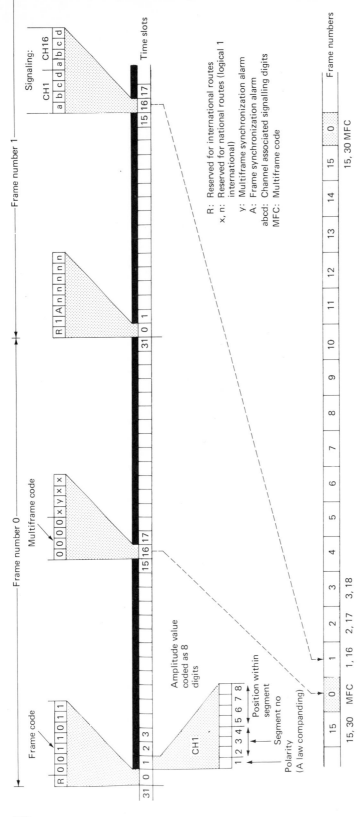

**Fig. 6-16** The frame format of the European 30-channel multiplex system. (This format is also used in Latin America, Australasia, and Africa.)

Digits 5, 6, 7, and 8    Indicates the sample's position within the segment

The channel messages are combined by the process of word interleaving, the 30 words of 8 digits each, into a frame consisting of 32 time slots, numbered 0 to 31. The two spare time slots (0 and 16) are allocated to the functions of frame alignment and signaling as shown in the diagram.

A multiframe structure is required in order that channel associated signaling may be communicated. There are 16 frames per multiframe, which are conventionally numbered 0 to 15 (see Fig. 6-16).

The FAW appears as a 7-digit word occurring within time slot 0 in frames 0, 2, 4, 6, and so on. The remaining digit $R$ within these time slots is reserved for monitoring and control applications over international routes.

The MFAW appears as a 4-digit word occurring within time slot 16 during frame 0 only. The remaining 4 digits (time slot 16, frame 0) are reserved for housekeeping functions for both national and international routes. During all other frames time slot 16 is used for conveying channel associated signaling information (frames 1 to 15).

Multiplexers of this type[7] are currently used extensively within Europe, Africa, and South America.

## 6.6  TIME DIVISION MULTIPLE ACCESS SYSTEMS

It is unlikely within any communication system that every TDM tributary is simultaneously carrying information messages. This inefficiency may be reduced by specifying a multiplexing scheme where the number of tributary inputs exceeds the number of output channels available. Multiplex structures of this type are known as *time division multiple access* (TDMA) systems, and are frequently used in conjunction with satellite links.[8]

The use of TDMA equipment demands that the statistics of the message signals is taken into account such that there is always a very high probability that the number of output channels will not exceed the number available. In the case of TDMAs used for speech messages, overload conditions are permissible provided that they are of such a short duration that they can be ignored by the human ear.

It is important in the case of data and telex communication that no information loss occurs, while the effect of a transmission delay may be ignored. Consequently, such signals are typically accumulated at the terminals and transmitted as complete messages, or as fixed-length sequences known as *packets.* Data transmission networks exist today based on both techniques and are referred to as *message-switching,* or *packet-switching* systems, respectively. The TDMA equipment used for data messages is frequently referred to as a *concentrator,* although this term is usually reserved for describing general switching machines.

A further increase in the number of speech channels that may be accommodated within a given system can be obtained by using *time-assignment speech interpolation*

---

[7] International Telegraph and Telephone Consultative Committee, *CCITT Orange Book,* vol. III-2, recommendations G711 and G732.

[8] G. C. Hall, "Single Channel per Carrier, Pulse Code Modulation, Demand Assignment Equipment (SPADE) for Satellite Communications," *Post Office Electrical Engineers' Journal,* vol. 67, pp. 42–48, 1974.

(TASI) techniques.   In a two-way conversation it is clear that each observer is talking for less than half the time on average.   TASI equipment uses speech detectors and high-speed electronic switches to connect other users when any given route contains no speech signals.   When the original observer resumes speaking, a new communication path is provided.

# 7

# Asynchronous Time
# Division Multiplexing

Our concern in this chapter will be the technology involved in combining digital information streams that have been derived from different clock sources. Such channel signals are referred to as *asynchronous* since their significant instants do not necessarily coincide.

We shall see that a process known as *justification,* or *pulse stuffing* must be used to synchronize each of the channel signals to a common clock reference. Once this intermediate process has been performed, the channel signals may be synchronously multiplexed as described in the previous chapter.

Consequently, we shall describe here mainly the technique of synchronizing asynchronous signals to a common reference clock. In this context we shall consider various frame formats that give rise to purely positive, purely negative, and positive-negative justification systems. The merits and limitations afforded by the different frame formats will be identified.

We shall adopt as an example (see Fig. 7-1) the European requirement for an asynchronous second-order TDM that combines 4 asynchronous channels each of bit rate = 2.048 Mbit/s ($\pm$ 50 ppm) to yield an aggregate signal of 8.448 Mbit/s ($\pm$ 30 ppm). The CCITT[1] recommends a suitable frame format for this multiplex and we shall conclude with a description of this recommendation. However, first we shall consider several other simplistic formats that would also satisfy the design requirement.

## 7.1 BASIC IDEAS

The problem of combining 4 asynchronous channel signals at 2.048 Mbit/s to form an aggregate signal of 8.448 Mbit/s is best investigated at the multiplexed channel rate, that is 2.112 Mbit/s. Our aim will be to define a frame format such that 2.048 Mbit/s of information digits, plus some control digits, including those allocated for frame alignment purposes, may be contained within the 2.112 Mbit/s stream. In this case, provided that the channel and multiplex clocks are absolutely accurate, or are synchronized to each other, the ratio of the information-to-control digits $n$ is given by

[1] International Telegraph and Telephone Consultative Committee, *CCITT Orange Book,* vol. III-2, G742 (recommendation), International Telecommunications Union, Geneva, Switzerland, 1977.

$$n = \frac{2.048}{2.112 - 2.048} = 32:1 \tag{7.1}$$

Using this result we may propose a frame code that allocates 2 consecutive digits for control purposes, followed by 64 information carrying digits. The resultant frame format is illustrated within Fig. 7-2. The multiplexed signal $b$ has been included for completeness, and to draw attention to the fact that a frame alignment word of 8 digits can be formed by synchronous bit interleaving of the four tributary signals defined by Fig. 7-2$a$.

It is appropriate to consider the mechanism required to insert the control digits within the 2.112 Mbit/s information stream. A circular store is used to cyclically write the channel information digits into the various storage locations at a rate of 2.048 Mbit/s. The store is systematically read at a rate of 2.112 Mbit/s, but stops periodically in order to insert the 2 control digits per 64 information digits. The arrangement is illustrated within Fig. 7-3, together with a graphic representation of the phase relationship between the writing and reading clocks.

The diagram shows store location $A$ being written at the same instant as location $G$ is read. Now since the instantaneous reading rate (2.112 Mbit/s) is somewhat higher than the writing rate (2.048 Mbit/s), the system moves to the situation where location $A$ is written, while $H$ is read. The limiting condition occurs when location $A$ is written while $I$ is read. At this point 64 information digits have been transmitted, and the read clock is stopped for an interval lasting 2 digits; thereafter the cycle repeats itself.

The phase relationship that exists between the two clocks will be important in

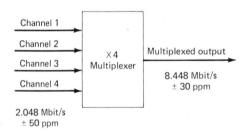

**Fig. 7-1** The European second-order asynchronous multiplexer.

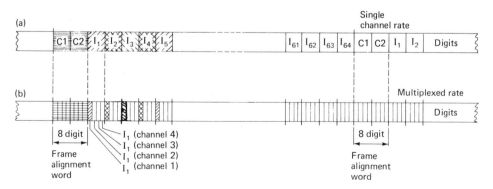

**Fig. 7-2** A simplistic frame format (second-order multiplexer).

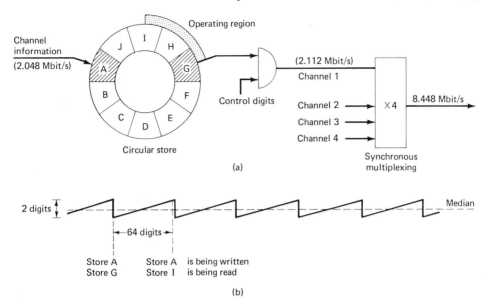

Fig. 7-3   (a) Insertion of justification control digits, using a cicular store; (b) the phase relationship between the write and read clocks.

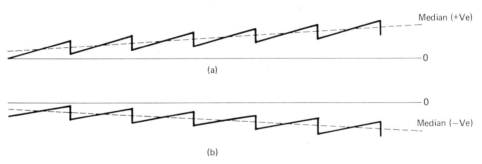

Fig. 7-4   The phase relationship between the write and read clocks: (a) write clock faster then read; (b) write clock slower than read.

the later sections of this chapter and needs further discussion.  If, as we have previously assumed, the system clocks are perfectly accurate then the average writing and reading rates of the store are equal.  This is signified by the median of the sawtooth phase relationship remaining level and constant.  Clearly this criterion must always be obeyed since if the mean writing and reading rates differ, information will be lost from the store.

Let us now allow the writing clock to have a slight inaccuracy, while the reading clock is maintained with zero tolerance.  (This assumption is valid since any inaccuracy in the reading rate may be considered as an increased inaccuracy of the writing rate.)  The degree of inaccuracy may be the previously assumed ±50 ppm, representing a positive or negative drift of order 100 Hz from the nominal writing rate of 2.048 MHz.  In this case the inherent stability between the writing and reading rates has been destroyed, and the median of the sawtooth phase relationship no longer remains constant (see Fig. 7-4).  In quantitative terms the average writing and reading rates of the store can no longer be assumed equal. A process known as *positive-negative justification* must be used to overcome the problem.

### 7.1.1 Positive-Negative Justification

The frame format adopted for positive-negative justification ($+/-$ J) must accommodate drifts of either polarity and typically has a form similar to that shown in Fig. 7-5.

The ratio of control-to-information digits $n$ has once more been maintained at 32, although the frame length has been increased threefold, and the control digits $J1$, $J2$, $X$, and $Y$ have been added.

The digits $J1$ and $J2$ are referred to as *justification control bits*. They signify that the clocks have drifted past a predetermined threshold (see Sec. 7.1.2), and the polarity of the drift. We may, for example, define the logical states of $J1$ and $J2$ in accordance with Table 7-1, where the action to be taken when a drift occurs is also identified.

Let us suppose that the write clock is operating too fast in comparison to the read. Since we have no control over the channel input information our only solution must be to speed up the reading process. This is achieved by reading the store during a control time slot $Y$, in addition to the normal information time slots within the frame. Thus time slot $Y$ may contain an information digit which must be extracted at the demultiplexer terminal. The frame is in this case said to be negatively justified, and is identified as such by the code 10 defined by the $J1$ and $J2$ digits.

The converse situation that occurs when the write clock is operating too slow in comparison to the read requires a frame that is positively justified. In this case

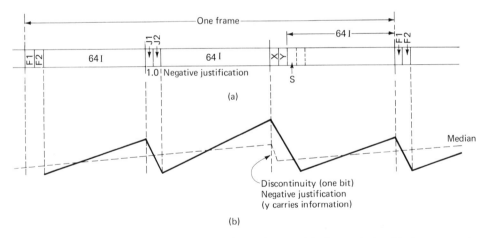

(a)

(b)

**Fig. 7-5** (a) Positive-negative justification frame format; (b) phase relationship between write and read clock during negative justification.

**Table 7-1  Justification Code Possibilities for Positive-Negative Justification**

| Write clock rate compared to read clock rate | Justification control bits | | Information carried by $Y$ time slot | Information carried by $S$ time slot | Justification type |
|---|---|---|---|---|---|
| | $J1$ | $J2$ | | | |
| Too fast | 1 | 0 | Yes | Yes | Negative $-Ve^J$ |
| Too slow | 0 | 1 | No | No | Positive $+Ve^J$ |
| Equivalent | 0 | 0 | No | Yes | None |

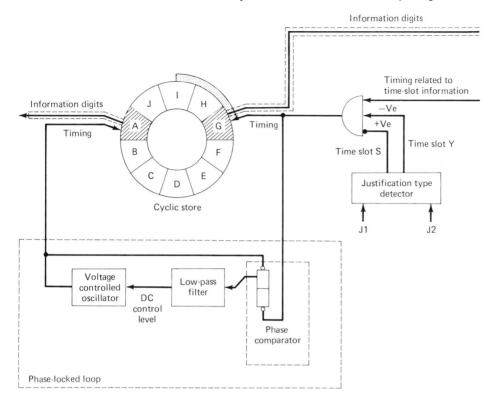

**Fig. 7-6**   Retrieving the original timing signal within the dejustification circuitry by using a phase-locked loop.

the reading process must be slowed down slightly, by, for example, not reading the store at time slot *S*.   Thus the total number of information carrying time slots within the frame is reduced by one.   Time slot *S* is conventionally referred to as a *justification,* or *stuffed* digit.   The code 01, defined by the digits *J*1 and *J*2 identifies the frame as positively justified.

If no drift occurs in either direction it is clear that no correction need be made to the reading rate.   The frame is identified as not being justified by the code 00, as shown in Table 7-1.

It is instructive to reexamine Fig. 7-5 in order to analyze the effect of justification on the phase relationship between the writing and reading clocks.   The correction introduced by either type of justification introduces a discontinuity within the median of the sawtooth phase relationship.   This ensures that the long-term average level of the median remains constant, and thus the writing and reading rates of the store are nominally equal.

The process of dejustification is essentially the reverse of that just described. In this case, however, it is necessary to detect not whether justification is required, but whether it took place at the multiplexer, as signified by the digits *J*1 and *J*2. We may consider that the dejustification circuitry performs two related functions: (1) retrieves the information digits from the designated time slots (including *Y* for $-Ve^J$; ignoring *S* for $+Ve^J$), and (2) regenerates a continuous clock signal with a rate nominally equivalent to the original channel information timing.

The clock signal must be regenerated by using a phase-locked loop (PLL) as shown in Fig. 7-6, where the timing reference signal supplied to the PLL is equivalent to

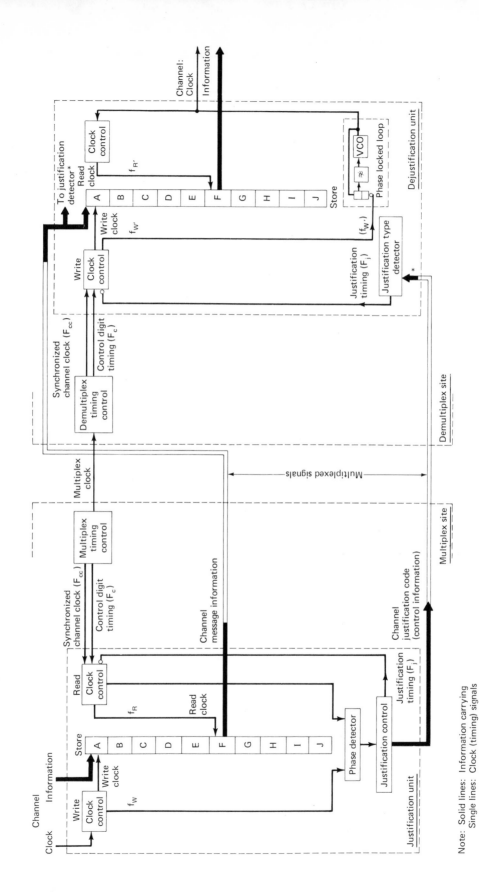

**Fig. 7-7** Block schematic showing timing information transfer during justification.

Note: Solid lines: Information carrying
Single lines: Clock (timing) signals

the channel message information rate. The resultant control signal supplied to the voltage controlled oscillator (VCO) is identical to the median of the sawtooth phase relationship identified earlier (Fig. 7-5); and consequently once more we may take note that since the long-term average level of the median remains constant the writing and reading rates of the store will be nominally equal. However, the instantaneous clock rate supplied by the PLL will vary, dependent on the prevailing level of the median.

The short-term variations of the regenerated timing signal from the ideal clock rate are known as *multiplex jitter*. [The term *waiting-time jitter* is also used. However, this is not completely accurate (see Sec. 7.3).] The major contribution to this jitter is due to the discontinuity within the median used as a control signal to the PLL, and cannot be avoided. The design of the PLL is quite critical also, and should not introduce drifts on its own account. For this reason the VCO requires a timing element that exhibits a high degree of short-term stability; typically a crystal is used. We shall consider the problems associated with the PLL design (Sec. 7.4), and multiplex jitter (Sec. 7.3) in more detail later.

At this point it is appropriate to recapitulate how the transfer of clocking information takes place within a justification system, and to ignore those details relating to the transfer of message information, which may be treated separately. The relevant circuit blocks are identified within Fig. 7-7.

Let us suppose that the channel message is clocked into the cyclic justification store at a rate $f_W$ bit/s. Then we may suppose that if the writing $f_W$ and reading $f_R$ rates at the store are to remain nominally equal, the following expression is valid:

$$f_W = f_R = F_{cc} - F_c \pm F_j \qquad (7.2)$$

where $F_{cc}$ = multiplex channel clock rate
$\quad F_c$ = control digit timing rate
$\quad \pm F_j$ = justification timing rate

A similar analysis at the dejustification store produces the expression:

$$f'_W = f'_R = F_{cc} - F_c \pm F_j \qquad (7.3)$$

where $f'_W$ and $f'_R$ represent the writing and reading rates, respectively, at the dejustification store.

The Eqs. (7.2) and (7.3) demonstrate that the justification process may be considered as a mechanism for conveying the timing error signal that exists between the write $f_W$ and the unjustified read $F_{cc} - F_c$ clocks. Furthermore the dejustified write clock $f'_W$ may be simply obtained if the value and polarity of the error signal $F_j$ is known.

The justification process occurs once each frame only, and we have considered in our example that a single digit of timing correction per frame will be sufficient to accommodate the clock inaccuracies $F_j$. In this case the maximum justification rate $F_{j\,\text{max}}$ equals the frame rate.

### 7.1.2    Justification Codes

We shall now consider the effect of an error within the justification code ($J1$ and $J2$). When this occurs the justification information will be incorrect, and an extra digit will be inserted, or removed, from the tributary information stream. Thus the number of bits within the tributary frame will be modified. This results in a loss of synchronism at all dependent demultiplexes.

Practical equipments must provide some immunity against justification errors, and in particular must be resistant to occasional burst errors. For this reason, typically, a distributed triplicated justification code is used, as shown in Fig. 7-8. Consequently any single block may be in error without affecting the justification information transfer.

In this case, if the probability of an error in any given bit is $P_e$, then the error rate for the justification control $P_j$ is given by (see Sec. 8.4.2):

$$P_J = 3P_e^2 \qquad (7.4)$$

### 7.1.3 Justification Phase Threshold

The moment when justification becomes necessary is defined by the justification phase threshold shown in Fig. 7-9. A separate threshold is required for each polarity of justification.

Let us consider that the justification mechanism is defined by Fig. 7-10. In this example the nominal operating region of the system is such that when store location $A$ is written, one of the stores located in region $X$ is being, or is about to be read. Now positive justification becomes necessary when the reading clock has advanced by 1 digit compared to the writing clock, as shown by state Y, Fig. 7-10. Thus an artificial threshold has been defined such that if store $A$ is written at the same instant as store $H$ is read, then justification must take place at the next opportunity.

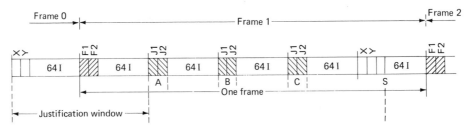

**Fig. 7-8** Specification of the justification window. (Note the triplicated justification code *A, B, C*.)

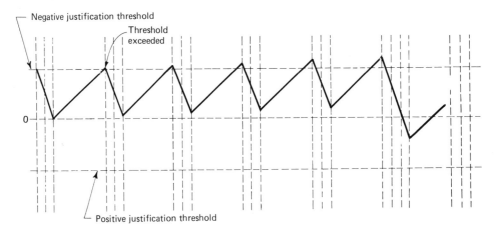

**Fig. 7-9** The justification phase threshold and the phase relationship between the write and read clocks.

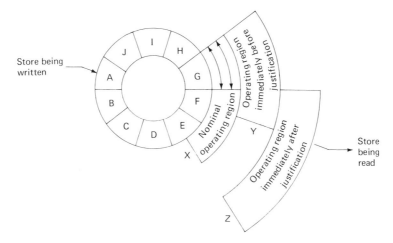

**Fig. 7-10** The phase relationship between write and read clocks within the justification store, viewed over a period of time.

It is important to realize that justification cannot be allowed to occur as soon as a phase threshold has been overtaken. The appropriate code $J1$ and $J2$ must always be transmitted (in triplicate[2]) prior to the act of justification, and for this reason the following operations must take place in the order given:

1. Detect that a phase threshold has been crossed
2. Transmit the triplicated justification code
3. Justify

Essentially this requires that a *justification window* must be defined (see Fig. 7-8). The phase relationship between the writing and reading clocks is examined during the time interval of the window only, since if the threshold is exceeded at any other moment justification must be delayed until the next frame.

We have considered the positive phase threshold as defined by when store $A$ is written at the same instant as store $H$ is read (i.e., write $A$/read $H$). However, the following phase thresholds are equally valid, since the store operates cyclically.

| Positive phase | | Write/Read |
|:---:|:---:|:---:|
| 1 | = | $A$ / $H$ |
| 2 | = | $B$ / $I$ |
| 3 | = | $C$ / $J$ |
| 4 | = | $D$ / $A$ |
| 5 | = | $E$ / $B$ |
| 6 | = | $F$ / $C$ |
| 7 | = | $G$ / $D$ |
| 8 | = | $H$ / $E$ |
| 9 | = | $I$ / $F$ |
| 10 | = | $J$ / $G$ |

---

[2] Triplication of the justification code word is used within the CCITT second- and third-order multiplexes; quintruplication is used within the fourth order.

If all of the possible phase thresholds are used (in our example phases 1 to 10 inclusive), then the waiting-time jitter introduced is as predicted by classical theory (Fig. 7-15, see Sec. 7.3). On the other hand, the jitter may be minimized over the operating region of the multiplex by selecting combinations of the phases appropriately (e.g., phases 1 and 5; or phases 1, 4, 7, and 10). We shall discuss this in more detail within Sec. 7.3.1.

We have previously considered that the tributary signal has a constant bit rate of 2.048 Mbit/s ($\pm$ 50 ppm). This is untrue, since typically the tributary signal will be influenced by pattern-induced line jitter (see Chaps. 8 and 10), and will exhibit continual variations in phase. The maximum permitted phase variations, or maximum jitter acceptance, of the system is determined by the excess storage capacity of the cyclic store. Thus in addition to choosing a combination of phases that seek to reduce the waiting-time jitter, the designer must also maximize the useful excess storage capacity.

Let us suppose that the waiting-time jitter introduced is minimized over the operating region of the multiplex (see Sec. 7.3.1) by choosing phases 1 and 5, given by

$$\text{Phase } 1 = \text{write } A/\text{read } H$$
$$\text{Phase } 5 = \text{write } E/\text{read } B$$

This, by reference to Fig. 7-10, gives us a maximum of approximately 3 bits ($H$, $I$, and $J$) excess storage in the positive direction. If the line jitter happens to have a magnitude of 4 bits, then location $A$ will be written and read simultaneously, and the storage capacity exceeded.

We may define Phases 1 and 5 differently such that:

$$\text{Phase } 1 = \text{write } A/\text{read } I$$
$$\text{Phase } 5 = \text{write } E/\text{read } C$$

In this case although the waiting-time jitter introduced remains the same as before, one bit of excess storage, in the positive direction, has been lost.

The negative justification phase threshold is determined using identical arguments. However, the excess storage provided in the positive and negative directions should be made equal since line jitter is always symmetrical.


## 7.2 FRAME FORMAT MANIPULATION

Previously we considered a frame format that requires no justification when the system clocks are perfectly accurate, or are synchronized to each other. In this case $n$ equals 32, for our chosen example [see Eq. (7.1)].

We shall now extend our analysis by choosing $n = 31$ and 33. In this case the writing and reading rates at the cyclic store can never be equal, unless the read clock is justified in the appropriate direction. Even if the system clocks are perfectly accurate, a certain amount of *fixed justification* will always be required.

Let us suppose that a given value of $n$ has a fixed positive justification rate of +3000 Hz, and that the inaccuracies in the system clocks amount to a total error of $\pm150$ Hz with respect to the nominal frequencies. In this case the amount of justification required will lie in the range 2850 to 3150 Hz and will always be in the positive direction. This simplifies the equipment design considerably, since negative justification circuitry may be omitted. The correction required for the error frequency is known as *variable justification*.

It is appropriate to analyze the case when $n = 31$, and 32, quantitatively.

## 7.2.1  Negative Justification ($n = 31$)

Let us assume a frame format as shown in Fig. 7-11, where $n$, the ratio of information to control digits, equals $31 : 1$.

The channel clock, or write clock rate $f_W = 2.048$ MHz. The multiplex channel read clock $F_R = 2.112 \times 31/32$ MHz. Therefore, the fixed justification rate $F_J$ is given by

$$F_J = F_R - f_W \tag{7.5}$$
$$= -2000 \text{ Hz} \tag{7.6}$$

The negative sign indicates that negative justification is required. Now, if the accuracy of $f_W$ is $\pm 50$ ppm, and $F_R$ is $\pm 30$ ppm, then the actual justification rate $J_A$ must be defined by the limits:

$$J_A = F_J \pm \text{ maximum error in } (f_W + F_R) \tag{7.7}$$

$$J_A = -2000 \pm \left( 2.048 \times 50 + 2.112 \times \frac{31}{32} \times 30 \right) \tag{7.8}$$

$$J_A = -2000 \pm (102.40 + 61.38) \tag{7.9}$$

$$J_A = -1836.22 \text{ Hz to } -2163.78 \text{ Hz} \tag{7.10}$$

However, by convention the actual justification rate is normalized by specifying it in terms of the maximum justification rate $F_{j\,\text{max}}$. Thus the normalized justification ratio $J_N$ is given by

$$J_N = \frac{J_A}{F_{j\,\text{max}}} = \frac{J_A}{\text{frame rate}} \tag{7.11}$$

where the frame rate $= \dfrac{\text{multiplex channel clock rate } F_{cc}}{\text{number of bits per frame}}$

$$= \frac{2.112}{320} = 6.60 \text{ kHz} \tag{7.12}$$

By substitution of Eqs. (7.10) and (7.12) into (7.11) we get:

$$J_N = -0.278 \text{ to } -0.328 \tag{7.13}$$

The parameter $J_N$ is used by the system designer as an indication of the proposed frame format's maximum theoretical jitter acceptance, and the likely value of the waiting-time jitter introduced by the equipment. We shall return to this point within Sec. 7.3.1.

The operating region of the multiplex is defined by the limiting values of $J_N$. It is this region that is important when designing the phase threshold detection mechanism.

## 7.2.2  Positive Justification ($n = 33$)

We shall now repeat the analysis performed in the preceding section choosing $n = 33$, and assuming the frame format shown in Fig. 7-12. In this case the fixed justification rate $F_J$ is given, by Eq. (7.5) as

$$F_J = \left( 2.112 \times \frac{33}{34} - 2.048 \right) \text{ MHz} \tag{7.14}$$

$$F_J = +1882.35 \text{ Hz} \tag{7.15}$$

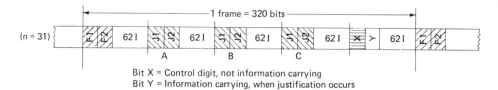

Bit X = Control digit, not information carrying
Bit Y = Information carrying, when justification occurs

**Fig. 7-11**  A frame format suitable for negative justification.

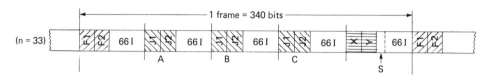

**Fig. 7-12**  A frame format suitable for positive justification.

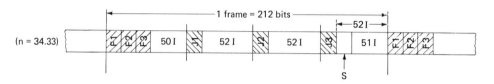

**Fig. 7-13**  An irregular frame format based on positive justification. The example shown is equivalent to that defined by the CCITT for the European second-order multiplex equipment (only channel level is shown here; a multiplexed signal is obtained by bit interleaving 4 channel signals).

The positive sign indicates that positive justification is required. Using the same clock tolerances as in the previous section, and by substitution into Eq. (7.7) gives the result:

$$J_A = +1882.35 \pm \left(2.048 \times 50 + 2.112 \times \frac{33}{34} \times 30\right) \text{Hz} \qquad (7.16)$$

$$J_A = +1718.45 \text{ to } +2046.25 \text{ Hz} \qquad (7.17)$$

The frame rate is given by substitution into Eq. (7.12) as

$$\text{Frame rate} = \frac{2.112}{340} = 6.21 \text{ kHz} \qquad (7.18)$$

Thus, the range of possible values for $J_N$ is given, by substitution of Eqs. (7.17) and (7.18) into Eq. (7.11), as

$$J_N = -0.276 \text{ to } +0.329 \qquad (7.19)$$

### 7.2.3  Irregular Frame Formats ($n = 34.333$)

Once again we shall repeat the above analysis in order to obtain the range of possible values for $J_N$. In this case, however, we shall consider the irregular frame format defined by Fig. 7-13, where $n$ is given by

$$n = \frac{\text{number of information bits in the frame}}{\text{number of control bits in the frame}} \qquad (7.20)$$

$$n = \frac{206}{6} = 34.333 \tag{7.21}$$

It should be noted that the frame is not completely irregular since there are an equal number of bits per subframe (53 bits). Furthermore the control digit $J_1$, indicating negative justification, has been deleted since it serves no useful purpose; yet the number of digits allocated to frame synchronization has increased (3 per channel).

The fixed justification rate $F_J$ is given by Eq. (7.5) as

$$F_J = \left( 2.112 \times \frac{34.333}{35.333} - 2.048 \right) \text{MHz} \tag{7.22}$$

$$F_J = -4226.41 \text{ Hz} \tag{7.23}$$

The positive sign, indicating positive justification, and the large value of fixed justification required, allows us to ignore negative justification entirely.

Using the same clock tolerances as in Sec. 7.2.1, and by substitution into Eq. (7.7) gives the result:

$$J_A = +4226.41 \pm \left( 2.048 \times 50 + 2.112 \times \frac{34.333}{35.333} \times 30 \right) \tag{7.24}$$

$$J_A = +4062.44 \text{ to } +4390.38 \text{ Hz} \tag{7.25}$$

The frame rate is given by substitution into Eq. (7.12) as

$$\text{Frame rate} = \frac{2.112}{212} = 9.962 \text{ kHz} \tag{7.26}$$

Thus the range of possible values for $J_N$ is given by substitution of Eqs. (7.25) and (7.26) into Eq. (7.11) as

$$J_N = +0.41 \text{ to } -0.44 \tag{7.27}$$

We shall see later (Sec. 7.3.1) that the operating region of a multiplex, defined by $J_N$, is capable of accepting a large amount of line jitter if it is located near to 0.5. Furthermore, the amount of waiting-time jitter introduced is minimized if the operating limits do not include the exact $J_N$ values 0.5, 0.3333, and 0.25. Thus, the frame format defined by Fig. 7-13 is very satisfactory from the point of view of jitter. A detailed analysis shows that the mean synchronization time, determined using Eq. (6.6), is very short and that the degree of irregularity of the frame is small. Therefore equipment implementation is not difficult. For these reasons this frame format has been recommended by the CCITT[3] for multiplexing four primary level signals to provide an aggregate signal of 8.448 Mbit/s.

## 7.3 MULTIPLEX JITTER

We have already described the mechanism used to transfer clock timing information from the tributary input at the multiplexer to the corresponding output port at the demultiplexer. We have seen that a PLL controlled by a waveform, such as shown in Fig. 7-5$b$, is used to regenerate a continuous clock signal with a rate nominally equivalent to the original tributary timing information.

---

[3] International Telegraph and Telephone Consultative Committee, *CCITT Orange Book,* vol. III-2, G742 (recommendation).

The clock regenerated by the PLL will gradually increase in frequency and then abruptly slow down again, due mainly to the process of justification. Consequently the resultant regenerated timing signal experiences small phase shifts compared to the original clock source. These phase shifts are referred to generally as *multiplex jitter*. However, more precisely it is possible to separate this jitter into the following three components:

1. A high-frequency contribution due to the periodic insertion of control digits (amplitude = 1 bit, frequency = 39.8 kHz; and also amplitude = 2 bits, frequency = 9.96 kHz).[4]
2. A component at the mean justification rate (amplitude = 1 bit, frequency = 4.2 kHz).[5]
3. A very-low frequency component, due to the fact that justification can occur only in a specific time slot each frame, and not as soon as it is required. This gives rise to the name *waiting-time jitter*.

The high-frequency systematic jitter contributions [(1) and (2)] are substantially removed by the low-pass transfer function of the PLL. Thus only the low-frequency beats that occur in the justification process (3), and cannot be removed by the PLL, are of concern.

Waiting-time jitter has an amplitude which is related to the normalized justification ratio $J_N$, and has spectral components that extend down to zero frequency. A theoretical analysis of waiting-time jitter has been made by several researchers, namely, Duttweiler,[6] Matsuura,[7] and Kuroyanagi,[8] and has resulted in rather complex mathematics. We shall consider here a simplified graphic treatment in an attempt to show more clearly how waiting-time jitter arises.

Let us suppose that a particular multiplexer has a value of $J_N$ equal to 0.3333 exactly. In this case justification occurs once in every three frames, and the phase relationship between the writing and reading clocks at the cyclic store is as shown in Fig. 7-14a.

Now small inaccuracies in the system clocks $\Delta$ may cause a slight variation in the value of $J$. This will result in a gradual drift in the phase relationship between the writing and reading clocks, and on occasion the justification phase threshold will be crossed at shorter (or longer) intervals, as shown in Figs. 7-14b and 7-14c. Eventually the phase threshold will be crossed sufficiently early such that the next justification opportunity may be used. In this case justification occurs usually once in every three frames, and occasionally at two (Fig. 7-14b), or four (Fig. 7-14c) frame intervals. This effect causes a low-frequency beat signal to be superimposed on the high-frequency jitter.

We may deduce from the above argument that if the value of $J_N$ is a simple fraction, such as $\frac{1}{4}$, $\frac{1}{3}$, $\frac{1}{5}$, $\frac{2}{3}$, $\frac{2}{5}$, $\frac{3}{5}$, etc., no waiting-time jitter will occur. In practice,

---

[4] Figures quoted relate to the CCITT G742 multiplex, and are given as an example only. For this example, the frame rate is 9.96 kHz, and the code insertion rate (subframe rate) is 39.8 kHz.

[5] Ibid.

[6] D. L. Duttweiler, "Waiting Time Jitter," *Bell System Technical Journal*, vol. 51, 1972, pp. 165–207.

[7] Y. Matsuura, S. Kozuka, and K. Yuki, "Jitter Characteristics of Pulse Stuffing Synchronization," *Proceedings of the Institute of Electrical and Electronics Engineers 1968 International Communications Conference*, Institute of Electrical and Electronics Engineers, 1968, pp. 259–264.

[8] N. Kuroyanagi, H. Saito, S. Kozuka, and Y. Okamato, *IEEE Transactions on Communications*, vol. COM-17, 1969, pp. 711–720.

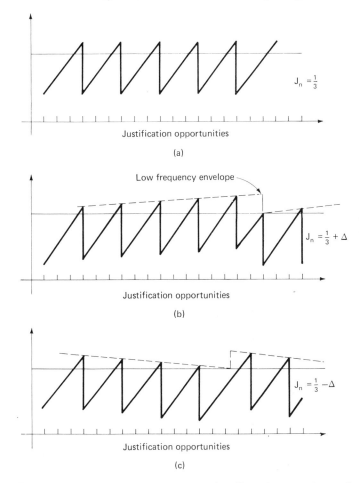

**Fig. 7-14**   The introduction of waiting-time jitter due to nonintegral justification ratios $J_N$.

however, due to inevitable inaccuracies in the system clocks $J_N$ will deviate slightly and give rise to a jitter amplitude $A_J$ of

$$A_J = J_N \qquad \text{bits} \tag{7.28}$$

It is instructive to plot $A_J$ against $J_N$ for the most significant jitter peaks. This has been done for the range $J_N$ equals 0 to +1 in Fig. 7-15; an identical plot is obtained in the negative justification direction (0 to −1). The point where $J_N$ equals zero, equivalent to positive-negative justification, contains a jitter peak of 1-digit amplitude.

We must take note that the jitter spectrum shown in Fig. 7-15 does not indicate the frequency of the jitter. The $x$ axis, that is the normalized justification ratio $J_N$, is dimensionless and is given by

$$J_N = \frac{\text{actual justification rate}}{\text{maximum justification rate}} \tag{7.29}$$

We have made two important assumptions about the justification phase threshold that need clarification.

First, we have considered that the phase threshold maintains a constant level. This assumption is invalid if the tributary write clock, at the cyclic store, is affected by line jitter on the input signal.

Second, we have assumed that the phase threshold is examined, and that the decision to justify can be made at all times during the justification window. For certain equipment implementations this will be untrue.

Now if the above assumptions are valid the classical jitter characteristic, as shown in Fig. 7-15, is obtained. Otherwise the characteristic will be dramatically modified.

### 7.3.1 Optimizing the Jitter Parameter

Let us assume the classical jitter characteristic shown in Fig. 7-15 and extend our arguments to real systems later.

By consulting Fig. 7-15 it is clear that if $J_N$ is very small (say, 0.1), or very large (say, 0.9), then the waiting-time jitter introduced will be low. However, in this case the variable justification capacity of the system will be reduced to a minimum also (1 opportunity per 10 frames). Thus limiting values of $J_N$ do not allow the multiplexer to accept jittery input signals. On the other hand if $J_N$ is centered at 0.5, the variable justification capacity of the system is maximized. However, in this case the waiting-time jitter introduced is at a maximum also.

A practical choice for $J_N$ lies between 0.4 and 0.5, where the most significant jitter peaks are avoided, and the variable justification capacity of the system is nearly maximum. These arguments led to the frame format described in Sec. 7.2.3 being recommended by the CCITT for the second-order multiplex. In this example $J_N$ occupies the range $+0.41$ to $+0.44$ [see Eq. (7.27)], as identified within Fig. 7-15.

It is interesting to note that there is good reason to specify system clock rates

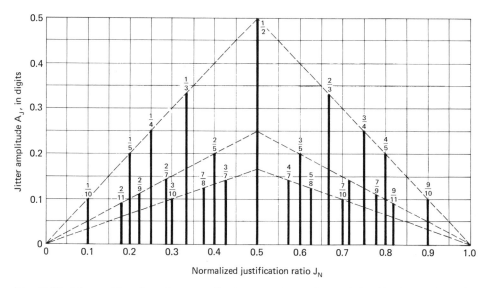

**Fig. 7-15** The waiting-time jitter amplitude occurring at integral justification ratios. The figure shows what is commonly referred to as the *classical* or *theoretical* waiting-time jitter characteristic.

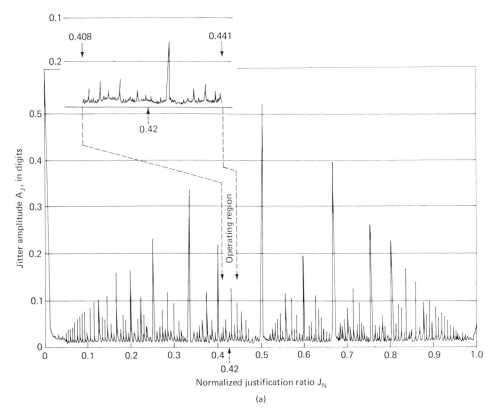

**Fig. 7-16a** Optimizing the jitter parameters: nonoptimized measured response. (From C. C. Cock, A. K. Edwards, and A. Jessop, "Timing Jitter in Digital Line Systems," *IEE Conference Publication no. 131, Telecommunication Transmission,* Sept. 1975. Copyright Standard Telephone Laboratories, Harlow, Essex, England.)

with close tolerances. This reduces the range of values that $J_N$ may pass through, and consequently minimizes the risk of hitting a jitter peak.

The equipment designer is not interested in the complete jitter characteristic that stretches from 0 to +1; but in the small portion defined by the jitter-free operating region of the multiplex. For example, the range 0.41 to 0.44, which is shown in Fig. 7-16.

The implementation of the justification phase threshold was discussed in Sec. 7.1.3, and is very important to the actual waiting-time jitter characteristic obtained within a given equipment. We noted that if all of the possible phases (phases 1 to 10; see Sec. 7.1.3) were used to define the justification phase threshold then the classical waiting-time jitter characteristic is recorded (see Fig. 7-16a). This is equivalent to examining the justification phase threshold at all times during the justification window: one of our original assumptions.

Alternatively we may decide to use only a combination of the available phases to define the threshold[9] (e.g., phases 1 and 5). In this case the recorded jitter characteristic will be different from the classical case and may offer an improvement over the operating region of the multiplex. A trial and error experimental process is normally

[9] British patent application, A. K. Edwards, no. 35606/73, 1973.

used to find the optimum combination of phases, since mathematical analysis is very complex.[10]  An optimized waiting-time jitter characteristic is shown for comparison purposes in Fig. 7-16*b*.

## 7.4  THE PHASE-LOCKED LOOP (PLL)

We have already noted that the function of the phase-locked loops, within the demultiplex tributary circuits, is to provide a continuous, jitter-free timing signal with a rate nominally equivalent to the original tributary input clock at the multiplex. The block schematic of the PLL and its function are illustrated within Fig. 7-6.

Normally the PLL employs a simple single pole low-pass filter within the loop. Such a filter has a roll-off characteristic of 20 dB per decade (equivalent to 6 dB per octave) as shown in Fig. 7-17.  Thus if the 3-dB point of the loop is one-hundredth of the frame rate, then jitter occurring at the frame rate will be attenuated by a factor 40 dB.  This is adequate for most purposes.

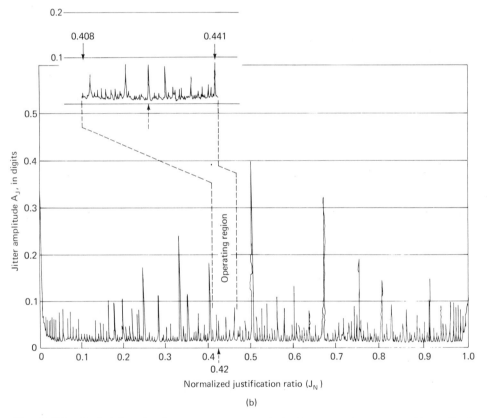

(b)

**Fig. 7-16*b***  Optimizing the jitter parameters: optimized measured response. (From C. C. Cock, A. K. Edwards, and A. Jessop, "Timing Jitter in Digital Line Systems," *IEE Conference Publication no. 131, Telecommunication Transmission,* Sept. 1975. Copyright Standard Telephone Laboratories, Harlow, Essex, England.)

[10] A theoretical treatment using a single phase is given by P. E. K. Chow, "Jitter Due to Pulse Stuffing Synchronization," *IEEE Transactions on Communication,* vol. COM-21, no. 7, 1973, pp. 854–859.

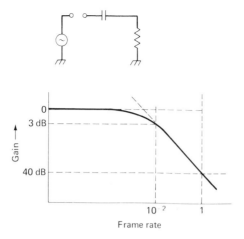

**Fig. 7-17**   Schematic, and response of a single pole low-pass filter.

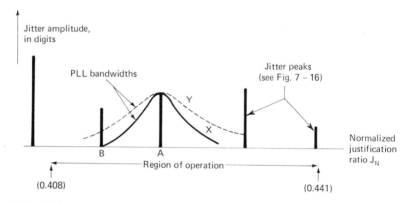

**Fig. 7-18**   Selecting the bandwidth of the phase-locked loop.

The CCITT second-order multiplex described in Sec. 7.2.3 has a frame rate $F$ of 9.96 kHz. Thus in this case the 3-dB loop bandwidth must be less than 100 Hz in order to reduce the high-frequency components of jitter to an insignificant level.

We have seen that the waiting-time jitter occurs at an extremely low frequency, which is dependent on the normalized justification ratio $J_N$. For this reason it is not possible to attenuate the maximum amplitude of any particular jitter peak (e.g., $A$ in Fig. 7-18) by decreasing the PLL bandwidth. However the PLL bandwidth should be sufficiently small such that only one of the most significant peaks (e.g., $A$), and not two ($A$ and $B$) occur within the pass band of the loop. Consequently a filter characteristic such as $X$ is required (Fig. 7-18). This criterion effectively reduces the average waiting-time jitter, provided that there is no relationship between the system clocks, and demands a loop bandwidth of order 40 Hz.[11]

Any line jitter that was present at the tributary input to the multiplex will be coupled to the PLL clock reference signal by the justification process. This coupling may be assumed ideal provided that the line jitter has a frequency much lower than

[11] Figures quoted relate to the CCITT G742 multiplex, and are given as an example only.

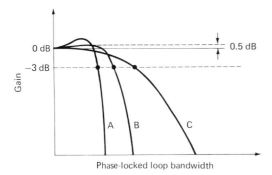

**Fig. 7-19** Typical gain/frequency characterization of a phase-locked loop.

the frame rate. Now, the presence of line jitter will modify the waiting-time jitter characteristic, and cause the broadening of the jitter peaks.[12] In order to remove the line jitter, and to limit the waiting-time jitter power at the PLL output, it is desirable to make the PLL bandwidth as low as possible.

Unfortunately if the PLL bandwidth is reduced to an exceptionally low value (e.g., 5 Hz) then other characteristics of the PLL will be adversely affected. The PLL may, for example, begin to amplify, rather than attenuate, the jitter at certain jitter frequencies (plot *A* in Fig. 7-19). This will cause problems in a telephone network, where it is likely that several multiplexers and their PLLs will operate in tandem. Consequently the maximum amplification permitted by each PLL at any possible jitter frequency must be kept very low. A maximum limit of 0.5-dB gain per PLL, and a 3-dB cutoff frequency of order 25 Hz,[13] may be regarded as adequate for most applications (plot *B* in Fig. 7-19).

The *pull-in time* of the PLL, defined as the time taken for the PLL to synchronize with its reference clock, will be long if the loop bandwidth is low. For this reason some equipments include special circuitry that are designed to speed up the pull-in action (see Chap. 10). However, in some instances a long pull-in time is actually desirable.

Occasionally, during periods of high transmission error rate, it is possible that the PLL will have no reference signal, and will start drifting towards its free-running frequency. In this case, if the PLL has a slow response it will continue to remain in quasi-synchronism with the reference signal for some time. Thus, if the absence of the reference timing is brief, there will be no noticeable effect on the regenerated tributary clock.

Clearly the maximum acceptable pull-in time is dependent on the actual application of the multiplex involved. In the case of higher-order multiplex equipments used within the telephone network, a value of less than 1 second (s) is reasonable.

It is important that even when the PLL loses its reference timing for long periods, it must continue to supply a nominally accurate tributary clock signal. (The CCITT primary multiplex (G711 and G733) and the Bell T1 system both require a clock accuracy of better than ±50 ppm.) This feature permits the tuned elements within

---

[12] Duttweiler, "Waiting Time Jitter."

[13] Figures quoted relate to the CCITT G742 multiplex, and are given as an example only. See also Secs. 10.4 and 10.6, where the problems of accumulated jitter within a complete digital transmission system are reviewed.

the transmission system to function normally. However, it adds a further complication to the design of the PLL, since an accurate and stable timing element, a quartz crystal, must be used within the VCO circuitry.

The inclusion of a quartz crystal within the VCO reduces the pull-in range of the loop considerably. We may deduce the minimum acceptable pull-in range in the following manner: Let the clock tolerance of the tributary input to the multiplex = $x$, ppm, and the clock tolerance of the free running (unlocked) PLL = $y$, ppm. Therefore,

$$\text{Minimum pull-in range} = x + y \qquad \text{ppm} \qquad (7.30)$$

Normally the tolerances $x$ and $y$ are made equal.

The requirements identified above that need to be imposed on the design of the PLL are quite severe. Furthermore the PLL application considered here is somewhat different from the normal usages to which PLLs are commonly applied. A mathematical treatment of the design principles for loops of this type has been made by Byrne.[14]

### 7.5  PHASE TRACKING

During recent years several schemes have been proposed that seek to improve the justification process in one way or another. The concept of transmitting, in addition to the justification code words, information about the phase relationship between the system clocks is one of these, and is described here.

We have noted that the reference signal to the VCO is a sawtooth that has a frequency and amplitude equivalent to the waiting-time jitter. Typically this frequency $f_{WT}$ will be of order 0 to 40 Hz, and stretches over several frames; while the peak-to-peak amplitude is related to the justification ratio $J_N$ (see Fig. 7-16).

Let us suppose that a PLL identical to that used within the dejustification circuitry (Fig. 7-7), is also placed at the justification unit. If this second PLL is referred to the justified read clock it will contain waiting-time jitter of frequency and amplitude equivalent to that seen at the dejustification unit. The described configuration is illustrated within Fig. 7-20.

Now if the VCO control signal $W_s$ is digitized it may be transmitted within a modified frame format as shown in Fig. 7-21. The digits $A1$ to $A4$ (inclusive) represent the analog value of the inverted VCO signal. At the receive terminal the original analog value is reconstructed and is then subtracted from the VCO control signal. In essence this is equivalent to subtracting the waiting-time jitter, and results in a remarkably stable waveform at the PLL output. This technique has been successfully[15] employed within clock synchronizers intended for data traffic applications.

### 7.6  ASYNCHRONOUS MULTIPLEXERS FOR DATA TRANSMISSION

Digital multiplex techniques are increasingly being adopted in the field of data and telex transmission, in preference to conventional FDM multiplexers. Telex signals have a very low bit rate (50, 60, or 75 bit/s) and data signals have repetition rates that are normally less than 9.6 kbit/s. Therefore it is very attractive to combine

---

[14] C. J. Byrne, "Properties and Design of the Phase Controlled Oscillator with a Sawtooth Comparator," *Bell System Technical Journal,* March 1962.

[15] See for example: BTMC Transmission Equipment, "Time Slot Access Unit," Bell Telephone Manufacturing Company S.A., Antwerp, Belgium. (BTMC is an ITT subsidiary.)

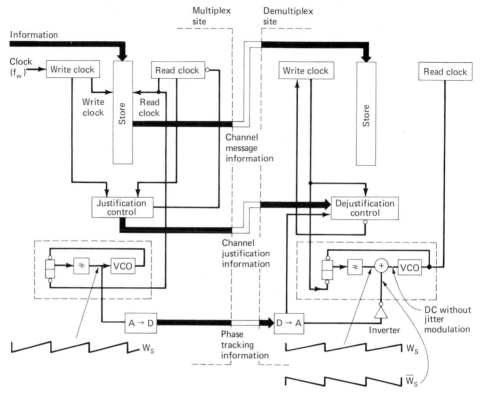

**Fig. 7-20** Jitter reduction based on the transmission of phase tracking data information. This diagram should be studied in conjunction with Fig. 7-7.

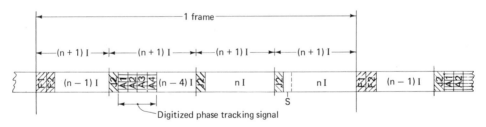

**Fig. 7-21** A modified positive justification frame format, incorporating a phase tracking signal.

several of these signals to form a higher bit rate aggregate message that may be more efficiently transmitted using existing facilities. Ideally several hundreds of telex signals and a suitable mix of data signals may be combined using TDMs to form an aggregate bit rate of 64 kbit/s. A signal at this rate may be economically transmitted using 1 channel of a PCM multiplex (8-kHz sampling, 8-bit coding), or alternatively by use of a group band modem.

Telex machines are normally synchronized to the electricity line frequency (60 Hz in the U.S.; 50 Hz in Europe), and must be assumed asynchronous with the TDM transmission clocks. Thus a *speed adapter* (Fig. 7-22) is required to synchronize the telex information to the TDM tributary clock rate. In such cases justification is not normally used since it is costly to implement on a per telex channel basis.

Less elegant, but nevertheless satisfactory solutions are employed, and these are described[16] in Secs. 7.6.1, 7.6.2, and 7.6.3.

In future years it is likely that all data sources will be synchronous with a single clock reference located within the telephone network. The concept of a synchronous data network will lead to simplified TDM equipments and switching machines. The same is true for a quasi-synchronous network that uses several free-running clock references that are kept within very close tolerances (e.g., $10^{-10}$ to $10^{-12}$). In the latter case, bit losses or repeated bits occur so infrequently that they can be tolerated.

It is already realized that the transmission facilities that are used within a synchronous network are likely to be over asynchronous bearers that happen to be cheaply available. We may for example consider the network configuration depicted by Fig. 7-22. In this example several telex machines are connected to a synchronous TDM arrangement, via a *speed adapter* (s). Some data signal sources are likewise so adapted, while any synchronous data signals are multiplexed directly. The aggregate signal may now be transmitted using a group band modem, in which case the transmission remains synchronized to clock reference (3); or alternatively a PCM transmission medium may be used. If PCM transmission is employed the TDM signal will be

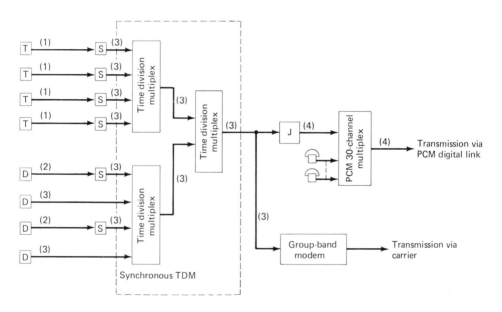

T  :  Telex machine, signal source

D  :  Data equipment, signal source

S  :  Speed adapter (e.g., Sampling—see Sec. 7.6.1)

J  :  Speed adaption, by justification (see Sec. 7.6.4)

(1), (2), (3), (4) : carriers are synchronized with respect to clock (1), (2), (3), (4)

**Fig. 7-22**  Asynchronous data and telex network configuration.

[16] H. C. Dinglinger, "Digital Multiplex Signals for Data Transmission," *ITT Electrical Communications,* vol. 52, no. 1, 1977.

conveyed using an asynchronous bearer (4), and positive-negative justification becomes necessary. We shall describe the asynchronous case in more detail within Sec. 7.6.4.

### 7.6.1  Clock Speed Adaptation by Sampling

By the sampling method, a data signal of bit rate $f_D$ may be sampled at a comparatively high rate $f_S$ to produce the desired synchronous equivalent data stream. The appropriate waveforms are illustrated within Fig. 7-23a.

The transition edges of the synchronized data signal will be distorted by the precision $D$ of the sampling process. The percentage sampling distortion $P_D$ is given by

$$P_D = \frac{f_D}{f_S} \times 100 \tag{7.31}$$

In most cases a distortion of less than 5 percent is acceptable. This corresponds to a sampling rate of 20 times the data signal bit rate, and leads to a very inefficient system. For this reason the sampling method is attractive when only a few low-bit-rate signals are to be multiplexed, and a suitable high transmission bit rate is available. This technique does, however, offer the advantage of a very cheap circuitry implementation.

### 7.6.2  Clock Speed Adaptation by Transition Encoding

The sampling method provides a transmission signal that has a large number of bits, which for the most part have an unchanging binary condition during the data bit interval. We shall consider here a method that seeks to reduce the redundancy within the transmitted signal by employing a suitable code structure.

The *sliding index* method describes the precise location of transitions, and the new logical value by coding, as shown in Fig. 7-23b. Information about the data message is, in the example, transmitted at one-quarter of the sampling rate.

Whenever a logical transition occurs the new binary condition is indicated first

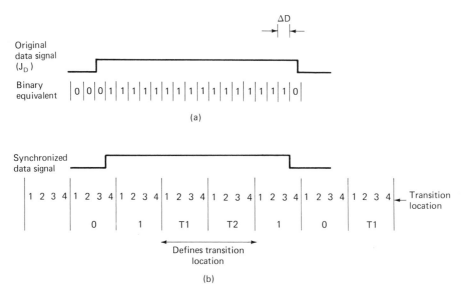

**Fig. 7-23**  Clock speed adaptation by (a) simple sampling (b) simple sampling followed by transition encoding.

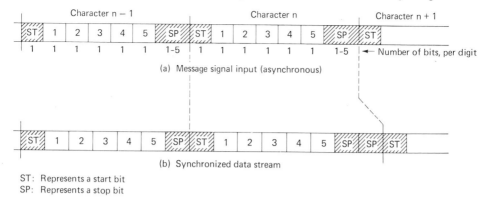

ST: Represents a start bit
SP: Represents a stop bit

**Fig. 7-24**   Clock speed adaption by start-stop bit adjustment.

(logical 1 in Fig. 7-23*b*).   Thereafter the next 2 digits (*T*1 and *T*2 in Fig. 7-23*b*) are used to define the exact sampling instant (1, 2, 3, or 4), where the transition occurred.   Thus a low transmission bit rate may be used in conjunction with a high sampling rate.

The above method permits a maximum data rate that is one-third of the transmission rate.   This limit is reached when each data transition can just be defined by a 3-digit sequence.

### 7.6.3   Clock Speed Adaptation by Start-Stop Bit Adjustment

Most data and telex signals do not require absolute code and speed transparency, provided that each multiplexed channel time slot is reserved for one type of message signal only.   For example, let us consider that only 50-baud (Bd) telex signals, as defined by CCITT alphabet no. 2,[17] are to be transmitted using a TDM.   In this case the telex signals will emit a continuous stream of characters[18] similar to those shown in Fig. 7-24*a*.   Each character, defined by a sequence of 5 bits, is preceded by a single start bit, and is terminated by a stop bit of duration 1.5 digits.   Therefore, the complete character length is 7.5 bits.

It is inconvenient in a bit-interleaved TDM structure to transmit half bits.   Therefore the sequence in Fig. 7-24*a* must be converted, prior to synchronous multiplexing,[19] to the form shown in Fig. 7-24*b*, where the stop bit may have a length of either 1 or 2 digits.   This readily permits a simple form of positive-negative justification.

When justification is unnecessary, adjacent sequences in the synchronized data stream will contain stop bits of different length (1 or 2 bits).   However, when justification is required then occasionally adjacent stop bits may be made equal in length (1 and 1, or 2 and 2 bits).   The adjustment in relative clock rates that may be obtained using this technique is clearly indicated in Fig. 7-24.

The original character interleaved sequence (Fig. 7-24*a*) may be readily regenerated without the need for transmitting additional justification coding information, since

---

[17] International Telegraph and Telephone Consultative Committee, *CCITT Orange Book*, vol. VII, recommendations S3 and S4, International Telecommunications Union, Geneva, Switzerland, 1977.

[18] Defined as: A letter, figure, punctuation, or other sign contained in a text to be transmitted by "alphabetic telegraphy" (CCITT definition 31.09).

[19] CCITT recommendation R101B, for a data and telex TDM, operating at 2.4 kbit/s.

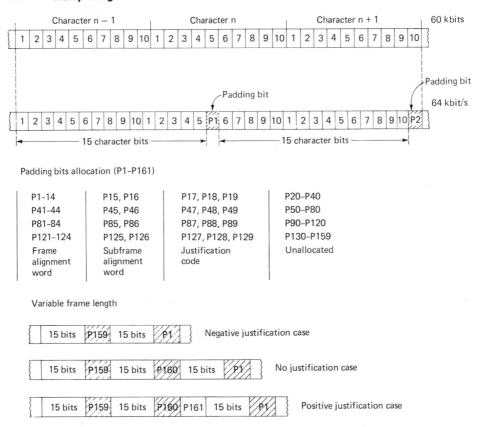

**Fig. 7-25**  Transmission over an asynchronous carrier using positive-negative justification.

the exact form of each 7.5-bit character sequence is known at the receive terminal where the characters are reconstructed.

This method of speed adaptation results in a very efficient utilization of the available transmission bandwidth.  Furthermore, unlike the transmission encoding method, there is no multiplication of transmission errors.  A single error in the transmission path results in a single error in the regenerated data signal.

### 7.6.4  Transmission over an Asynchronous Carrier

It is appropriate to relate our earlier discussions about positive-negative justification (Sec. 7.1.1) to the special problem of synchronizing a TDM signal for transmission using an asynchronous bearer.  We shall consider as an example the problem of conveying a 60 kbit/s character interleaved TDM signal[20] via a single PCM channel (64 kbit/s).  Let us further suppose that the integrity of each character (10 bits) must be preserved within the synchronized signal.

As a first step a padding digit may be inserted at 15-bit intervals, as shown in Fig. 7-25.  This effectively increases the bit rate by an amount $^{16}\!/_{15}$, in order to

[20] F. R. E. Dell, "Features of a Proposed Synchronous Data Network," *IEEE Transactions on Communications,* vol. COM-20, no. 3, 1972, pp. 499–503; and CCITT recommendation X51.

produce a signal with a repetition rate of 64 kbit/s. Now, justification may be simply performed by occasionally inserting 2 padding bits together (positive), or by omitting a padding bit (negative). The technique is illustrated in Fig. 7-25, where it is clear that the desired frame structure will have a variable length. When no justification occurs there will be 2560 bits per frame; otherwise the frame may stretch over 2559, or 2561 bits. This requires that the frame synchronization circuitry takes account of the justification condition and does not necessarily expect the FAW to occur at exact one-frame intervals.

The maximum speed adjustment SA that can be made using this technique is given by

$$SA = \pm \frac{\text{number of justification bits available, per frame}}{\text{number of bits, per frame}} \times \text{bit rate} \qquad (7.32)$$

$$= \pm \frac{1}{2560} \times 64 \text{ kHz} \qquad (7.33)$$

$$= \pm 25 \text{ Hz} \qquad (7.34)$$

Let us assume that the original 60 kbit/s data signal is synchronized to a synchronous data network, and that a highly accurate clock is used as the network reference. In this case the speed adjustment required will be almost entirely dependent on the clock accuracy of the asynchronous bearer. We have already noted that in the case of PCM a tolerance of ±50 ppm is permitted. This, related to the 64 kbit/s bearer, will demand a speed adjustment capacity $c$ of

$$c = \pm \frac{64}{1000} \times 50 \text{ Hz} \qquad (7.35)$$

$$c = \pm 3.2 \text{ Hz}$$

Consequently, the available speed adjustment using the proposed frame format is sufficient for communication via an asynchronous PCM bearer.

The justification of 1 bit per 2560 bits demands an exceptionally long frame, and gives rise to frame synchronization problems. For this reason each frame must be divided into four subframes that are separately identified by the padding bits $P15$–16, $P45$–46, $P85$–86, and $P125$–126. Each subframe starts with an identical 14-digit sequence that is used to obtain a rapid short-term synchronization time (see Chap. 6).

The method of speed adaptation by justification leads to a very efficient utilization of the available transmission bandwidth. However, more circuitry is required compared to all other methods, and for this reason it is only economically feasible for high-bit-rate data signals.

## 7.7 SELECTION OF A FRAME FORMAT

Within this chapter we have investigated the various problems that exist in the design of an asynchronous multiplexer. Furthermore we have demonstrated how the choice of frame format has a large influence on the system parameters. It is appropriate, then, to summarize the design approach employed in the selection of an optimum frame format for any particular requirement. The tasks are performed in the approximate order that they are listed below.

1. Determine an approximate value for the ratio of information-to-control digits *n*, using the formula

$$n = \frac{t}{m - t} \qquad (7.36)$$

   where *t* = input tributary bit rate
   *m* = multiplexed tributary bit rate

2. If the value of *n* is an integer, then it is possible to choose a positive-negative justification frame format. Normally this type of justification will only be selected for certain applications within synchronous networks, or where clock speed adaptation is required between two nominally equal bit rates. Asynchronous multiplexers based on the principle of positive-negative justification exhibit a high value of waiting-time jitter (peak-to-peak: 1 digit), and require a more complex circuitry implementation compared to a justification system of single polarity.

3. If the value of *n* has a fractional part, or alternatively if a frame format is chosen such that the calculated ratio of information-to-control digits is given by $n \pm x$, a certain amount of fixed justification $F_J$ will be required. The quantity *x* determines the values of $F_J$, and of the normalized justification ratio $J_N$, which may be calculated using Eq. (7.11). It can be assumed that justification of a single polarity will always occur provided $F_J$ is considerably larger than the variable amounts of justification required to accommodate the system clock tolerances. A system based on single-polarity justification simplifies the equipment design, and can be arranged to give low values of waiting-time jitter.

4. The value of the normalized justification ratio $J_N$ should ideally lie in the range 0.4 to 0.5, or 0.5 to 0.6. In this case the justification capacity of the system is maximized, while the most significant waiting-time jitter peaks are avoided (see Sec. 7.3.1). The above steps should be repeated until this criterion is met.

5. The frame synchronization time should be determined using Eq. (6.6). If necessary, the frame alignment word length, the percentage of digits given to synchronization, and the alignment strategy may be modified in accordance with the rules described within Chap. 6.

6. The number of digits reserved for coding the justification condition must be assessed on the basis of protection against occasional errors. The code word should be repeated at intervals to avoid disturbances due to burst errors.

7. Provision should be made for the communication of alarm and maintenance indications within the network. The time slots reserved for this purpose are conventionally referred to as *housekeeping bits* (see Chap. 6).

8. An irregular frame format often provides the best compromise between the parameters identified above. However, before such a format is adopted, the potential difficulties involved in the realization of the equipment should be explored.

The higher-order hierarchy adopted within Europe is based on multiplexers of very similar structure. The most important parameters of these equipments are listed in Table 7-2.

**Table 7-2  Frame Structures for Higher-Order Multiplex Systems (Europe)**

| System characteristics | Second order | Third order | Fourth order |
|---|---|---|---|
| Aggregate bit rate | 8.448 Mbit/s ± 30 ppm | 34.368 Mbit/s ± 20 ppm | 139.264 Mbit/s ± 15 ppm |
| Number of tributaries | 4 | 4 | 4 |
| Tributary bit rate | 2.048 Mbit/s ± 50 ppm | 8.448 Mbit/s ± 30 ppm | 34.368 Mbit/s ± 20 ppm |
| Type of justification | Purely positive | Purely positive | Purely positive |
| *Time-slot allocation* | *Subframe 1* | *Subframe 1* | *Subframe 1* |
| Frame alignment word | Bits  1–10 (1111010000) | Bits  1–10 (1111010000) | Bits  1–12 (111101000000) |
| Housekeeping bits | Bits 11, 12 | Bits 11, 12 | Bits 13, 14, 15, 16 |
| Bit-interleaved tributaries | Bits 13–212 | Bits 13–384 | Bits 17–488 |
| | *Subframes 2 and 3* | *Subframes 2 and 3* | *Subframes 2, 3, 4, and 5* |
| Justification control (one bit per tributary) | Bits 1–4 | Bits 1–4 | Bits 1–4 |
| Bit-interleaved tributaries | Bits 5–212 | Bits 5–384 | Bits 5–488 |
| | *Subframe 4* | *Subframe 4* | *Subframe 6* |
| Justification control (one bit per tributary) | Bits 1–4 | Bits 1–4 | Bits 1–4 |
| Justified digits (one bit per tributary) | Bits 5–8 | Bits 5–8 | Bits 5–8 |
| Bit-interleaved tributaries | Bits 9–212 | Bits 9–384 | Bits 9–488 |
| Frame length | 848 bits | 1536 bits | 2928 bits |
| Frame rate | 9.962 kHz | 22.375 kHz | 47.56 kHz |
| Normalized justification ratio $J_N$ (nominal value) | 0.424 | 0.436 | 0.419 |

**Table 7-3  Frame Structures for Higher-Order Multiplex Systems (North America)**

| System characteristics | Second order | Third order | Fourth order |
|---|---|---|---|
| Aggregate bit rate | 6.312 Mbit/s ± 30 ppm | 44.736 Mbit/s ± 20 ppm | 274.176 Mbit/s ± 10 ppm |
| Number of tributaries | 4 | 7 | 6 |
| Tributary bit rate | 1.544 Mbit/s ± 50 ppm | 6.312 Mbit/s ± 30 ppm | 44.736 Mbit/s ± 20 ppm |
| Type of justification | Purely positive | Purely positive | |
| *Time-slot allocation* | | | |
| Service digit | Bit 1 | Bit 1 | |
| Bit-interleaved tributaries | Bit 2–49 | Bit 2–89 | |
| Service digit sequence | $F_1, M_x, J_{TA}, F_0, J_{TB}, J_{TC}$ repeated cyclically for $x = 1$–$4$; $T = 1$–$4$ | $F_1, M_x, F_1, J_{TA}, F_0, J_{TA}, F_0, J_{TA}$ repeated cyclically for $x = 1$–$7$; $T = 1$–$7$ | |
| Subframe alignment | $F_1, F_0 = 10$ | $F_1, F_1, F_0, F_0 = 1100$ | |
| Multiframe alignment | $M_1, M_2, M_3, M_4 = 0111$ | $M_1, M_2, M_3, M_4, M_5, M_6, M_7 = 1110010$ | |
| Justification control | $J_{TA}, J_{TB}, J_{TC} =$ tributary $T$, $T = 1$ to $4$ | $J_{TA}, J_{TA} =$ tributary $T$, $T = 1$ to $7$ | |
| Justified digits | Immediately after $F_1$<br>Bit 2 = tributary 1<br>Bits 3–5 = tributary 2–4, respectively | Immediately after $F_1$ and prior to $M_x$<br>Bit 2 = tributary 1<br>Bits 3–8 = tributary 2–7, respectively | |
| Multiframe length | 1176 Bits | 4760 Bits | |
| Multiframe rate | 5.367 kHz | 9.368 kHz | |
| Normalized justification ratio $J_N$ (nominal value) | 0.334 | 0.390 | |

The North American hierarchy has been developed along different lines, although the principles remain identical to those described here.   The main difference between the two systems is that the North American scheme employs distributed frame alignment words, and that a multiframe is specified.   Furthermore, different numbers of tributaries are combined at each level in the hierarchy.   The most important parameters of the North American equipments are listed in Table 7-3.

# TRANSMISSION

# 8

# Transmission of
# Digital Signals

## 8.1 INTRODUCTION

The analysis of a digital transmission system can be quite complex, when considered as a complete, integrated unit. Consequently it is necessary to make certain simplifications, and to divide each transmission path (*A* to *B*) into smaller, easier to comprehend entities. The most obvious simplification that can be made is in terms of the equipments required to implement a given transmission path. A typical arrangement is illustrated within Fig. 8-1, where the following items have been identified:

- Line Terminal Apparatus (LTA)
- Repeater
- Transmission Medium

The design of any transmission system will be governed by the characteristics of the transmission medium employed. The other components within the system serve to counter the degradations of the transmitted signal caused by the transmission medium. For example, if we were to investigate the affect of attenuation, phase shift, and crosstalk on the transmission process, it would be clear that the digital signal must be periodically cleaned up during its journey along the transmission path. This operation is routinely performed by repeaters located at appropriate intervals along the route.

In this chapter we shall concern ourselves with describing the fundamental problems involved for low-bit-rate transmission using low-quality symmetric-pair cables, and high capacity systems that employ high-quality symmetric-pair, or coaxial cables.

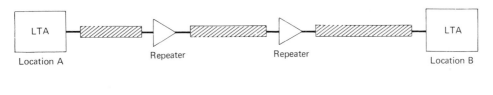

**Fig. 8-1** The main constituents of any transmission link *A* to *B*.

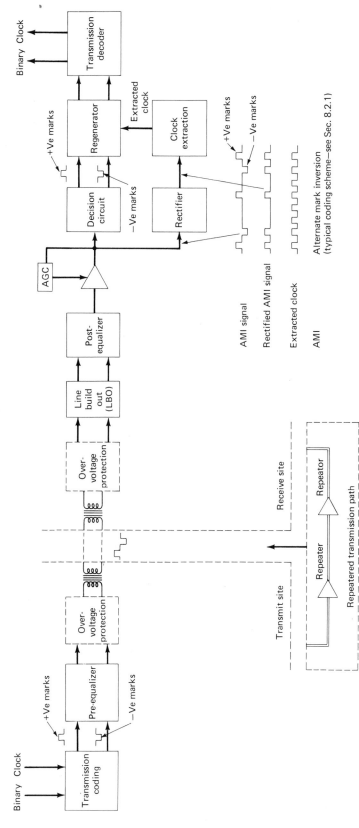

**Fig. 8-2** Block schematic of the line terminal apparatus: transmit and receive units.

These are the most common examples of digital transmission today, and their analysis will be helpful in understanding other, more modern, systems based on radio link, and optical fiber media.

We shall begin our description by subdividing the transmission equipments into still smaller component parts. This will enable us to understand the various requirements imposed by the transmission medium on the equipment. Furthermore we shall be able to illustrate the interactions that occur between the individual modules. We shall, however, refrain from going into too much detail at this time, saving this for the specialized chapters that follow (Chap. 9: "Transmission Codes;" Chap. 10: "Timing Extraction and Jitter;" and Chap. 11: "Equalization").

In the later sections of this chapter we shall concentrate on the practical problems that occur during the realization of a digital transmission system.

## 8.2 BASIC IDEAS

The functional blocks that make up a *line terminal apparatus* (LTA) or *regenerative repeater* are identified in Figs. 8-2, and 8-3, respectively. These are described below.

### 8.2.1 Transmission Coding

Binary signals cannot normally be transmitted over long distances directly. They must first be coded into a form that guarantees the following parameters:

1. A constant dc level along the transmission path
2. A suitable shaped energy spectrum
3. An adequate timing information content

Many coding structures offer additional desirable features such as the detection of errors, a reduction in the transmission bandwidth, and so on. These possibilities will be examined in Chap. 9.

The simplest transmission code that can satisfy the criteria identified above is the *alternate mark inversion* (AMI), or *bipolar* code. The conversion process interprets each binary (logical 1) pulse as a positive or negative *mark*, taken alternately, while the binary logical zero condition continues to be transmitted as such. The code translation is self-evident from Fig. 8-4.

The transmitted marks may occupy a full time slot (Fig. 8-4c), in which case they are referred to as *full baud,* or *nonreturn to zero* (NRZ) pulses. It is more usual however to employ *return to zero* (RZ) pulses as shown in Fig. 8-4d. The choice between the two schemes is dependent on the energy spectrum exhibited by the code and the consequent difficulty of extracting timing information (clock) from the transmitted line signal. Typically an equal mark/space ratio is used in the case of RZ transmission, although this is not absolutely necessary. (Optical systems based on laser emitters typically employ a 10 to 30 percent mark/space ratio to increase the lifetime of the laser.)

The energy spectra, assuming a random binary input signal, is shown for an AMI coded RZ and a binary line signal in Fig. 8-5. The chosen line code should ideally have a negligible energy at low frequencies; otherwise, physically large components will be required within the equalization circuitry. Furthermore, the attenuation introduced by the cable will be very small at low frequencies, and consequently overloading can occur at the repeater inputs if a significant energy level is permitted. Some

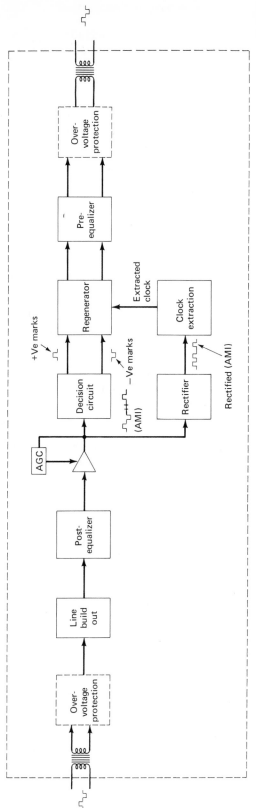

**Fig. 8-3** Block schematic of a digital-regenerative reader (cable transmission medium). The power feeding and supervision circuits have not been identified (see Secs. 8.7.1 and 8.7.3).

178

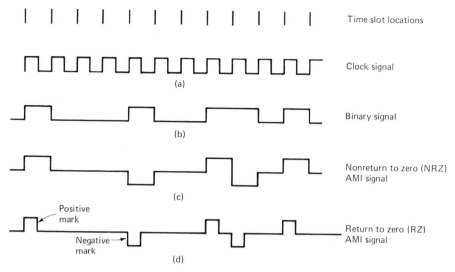

Fig. 8-4   Alternate mark inversion (AMI) coding of a binary plus clock signal.

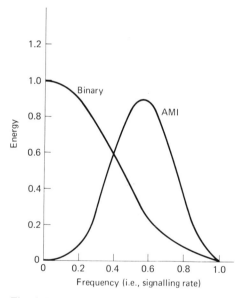

**Fig. 8-5**   Relative energy spectra of binary and
AMI signals.   Random signals are assumed in
both plots.

equipments overcome this problem by deliberately increasing the low-frequency attenuation by using a high-pass network.

The line code must be chosen such that a sufficient number of transitions, or
*zero crossings* occur.   These will be used to stimulate the timing information (clock)
extraction circuits.

The AMI code has an equal number of positive and negative marks, and consequently guarantees a constant dc level.   The absence of a dc component is important
for several practical reasons, which are listed below.

1. The decision threshold at which a mark is considered present or absent will ideally be linear. This cannot be so if the dc level is variable.
2. If an automatic gain control (AGC) is included within the repeater input circuitry, it will most likely use the pulse height as a control reference. A varying dc level causes the pulses to be displaced.
3. The transmission cable usually supplies a dc current to the dependent repeaters, in addition to carrying the message signal. In this case a constant dc level is most important.

The first two drawbacks can be overcome by adopting a more complex circuitry solution. However, the problem of power feeding (3) is much more difficult to solve.

The judicious choice of a transmission code is a most important part of any digital communication system. Distance, information rate, and the complexity of the circuitry required to implement the code will influence the selection. The following example demonstrates this fact.

Conventional practice within telephone exchanges is to keep multiplex and LTAs separate. Consequently, distances of up to 200 m between the multiplexer and LTA are not unknown. In this case a line code that is simple to implement, but requires a high transmission rate is satisfactory, e.g., high density bipolar 3 (HDB3). On the other hand, a more complex coding structure may be used if the information rate is high and larger transmission distances are involved [e.g., four binary, three ternary (4B3T)].

### 8.2.2  Timing Extraction

Repeaters and LTAs do not usually have their own built-in clock, but reconstruct the originally transmitted clock (see Fig. 8-4$a$) from the received line signal (see Fig. 8-4$d$).

Typically a resonant tank circuit, tuned to the peak energy of the coded line signal, is employed. The line signal must first be rectified to produce unipolarity marks, which are arranged to stimulate the tank circuit. The tank continues to oscillate, with an exponentially decaying amplitude, until the next mark stimulates the tank once more. This operation is illustrated within Fig. 8-6, where it is self-evident that the maximum distance between marks must not be too long. This criterion cannot be guaranteed using AMI coding, unless special conditions are imposed on the binary signal. This is a major disadvantage of the AMI code (see Chap. 9).

The resonant frequency $f_R$ of the tank will never be exactly equivalent to that of the originally transmitted clock $f_c$. Consequently the extracted timing signal will have a frequency $f_c$ at the moment of stimulation, and will thereafter drift toward a new frequency $f_R$. The extent of the drift will be dependent on the separation between marks, the values $f_R$ and $f_c$, and the $Q$ of the tank circuit. This effect results in an extracted clock that exhibits a continually varying phase, and is commonly known as *pattern-induced jitter* (see Chap. 10).

### 8.2.3  Regeneration

Let us consider the impairments that occur when a perfectly shaped pulse is transmitted over a cable. The pulse will be attenuated, dispersed, and subjected to the effects of random noise. In addition the effect of interference from other cables, that is *crosstalk,* must be taken into account.

Crosstalk and noise interference problems must be considered in relation to the

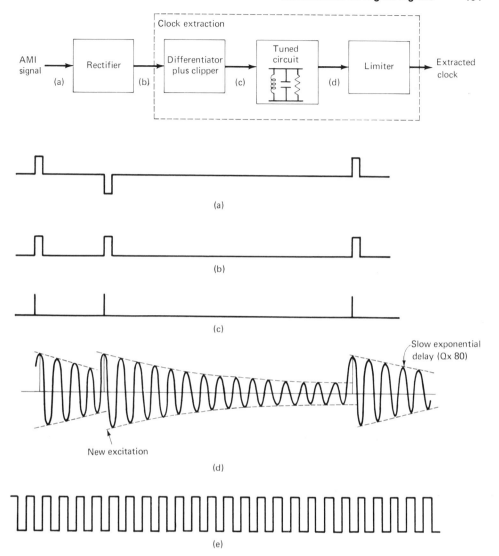

**Fig. 8-6** The function of the clock extraction circuit. The decay is emphasized (low Q) to illustrate the circuit operation.

pulse amplitude, which is in turn a function of the cable attenuation. The limiting situation largely determines the maximum acceptable repeater separation.

Pulse spreading, due to the effect of dispersion, will in the extreme cause intersymbol interference as shown in Fig. 8-7. Fortunately, however, digital signals are quantized both in amplitude and time; a fact that is exploited within the regeneration process by detecting the presence or absence of pulses at specified instants in time only. Consequently the level of the line signal outside the sampling instants may be conveniently ignored.

In theory the regeneration process can provide an almost error-free transmission performance. This ideal may be approached in practice by setting severe limits on the permitted amount of signal degradation before regeneration is deemed necessary.

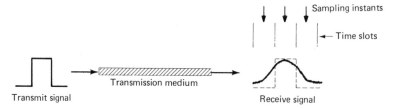

**Fig. 8-7**  Dispersion of the transmitted pulse by the transmission medium.

In this case it is necessary to consider the performance requirement of each regenerator section (i.e., regenerative repeater, plus the preceding length of cable) in isolation.

Let us suppose that the likelihood of a single error occurring in each regenerator section is $10^{-12}$, and that a particular transmission path comprises 50 sections. In this case the probability of an error for the complete system will be $50 \times 10^{-12}$ (see Sec. 8.4.5; this is an approximation). It is interesting to analyze the effect of a localized reduction in the transmission quality.

Let us assume that the possibility of an error increases from $10^{-12}$ to $10^{-7}$ in one of the sections. In this case the probability of an error within the complete system $P_S$ will be given by (see Sec. 8.4.5):

$$P_S = \text{no. of sections} \times \text{probability of error in each section} = nP_e \qquad (8.1)$$
$$P_S = 1 \times 10^{-7} + 49 \times 10^{-12} \qquad (8.2)$$
$$P_S \simeq 10^{-7} \qquad (8.3)$$

This example shows that the performance of a digital transmission system is dependent on the weakest link.

It is interesting to compare the above analysis to the case of analog FDM transmission. We shall consider here signal-to-noise ratio, rather than error probability, as a measure of analog transmission quality.

The signal-to-noise ratio $(S/N)$ and error probability $P_e$ can be shown, for gaussian noise, to be related by the curve defined in Fig. 8-13 (see Sec. 8.4.3).

The following relationship may be read directly from the curve:

$$
\begin{aligned}
P_e = 10^{-7} \qquad & (S/N)\text{dB} = 17.3 \text{ dB} \\
P_e = 10^{-12} \qquad & (S/N)\text{dB} = 19.9 \text{ dB}
\end{aligned}
\qquad (8.4)
$$

Thus in analog terms the considered degradation of the single section is equivalent to an increase in the noise power of 2.6 dB.

This additional contribution to the noise must be added to the total noise power for the 50 sections involved, in order to establish the effect on the system. Consequently we may consider a reduction in the transmission quality of one section as being shared between all others in the case of analog transmission.

### 8.2.4  Equalization

Equalization serves two purposes within a digital transmission system. In the first place it provides compensation for those cable parameters that are strongly frequency dependent. Secondly, it facilitates simple shaping of the transmitted pulses.

Coaxial and symmetric pair cables can be shown to exhibit the following frequency dependence:

$$\alpha = a + b\sqrt{F} + cF \qquad \text{dB/km} \qquad (8.5)$$

where $\alpha$ = attenuation, per unit length

$a$ = constant; typical value[1] = almost zero
$b$ = constant; typical value = 8.64
$c$ = constant; typical value = 0.05
$F$ = transmission frequency, MHz

$$\beta = \sqrt{LC}\,F \qquad \text{degrees of phase shift/km} \tag{8.6}$$

where $\beta$ = phase variation, per unit length
$L$ = characteristic impedance of the cable
$C$ = characteristic capacitance of the cable

The above equations are valid at high frequencies, when the ac and dc resistances become comparable. This typically occurs at frequencies above 200 kHz.

The value of constants $a$ and $c$ in Eq. (8.5) is very small. Consequently, unless very high transmission rates are involved, say above 100 MHz, it is possible to ignore the linear term. The attenuation in this case may be considered proportional to the square root of $F$.

Ideally the equalization circuitry will exhibit a frequency dependence which is the exact inverse of that recorded for the cable. In practice, however, it is necessary to compensate for variations of attenuation with frequency almost exactly, while providing only a limited amount of phase distortion correction.

The regenerator decision circuitry does not require a perfectly rectangular waveform; a near sinusoidal pulse shape is quite satisfactory. Such a pulse requires less bandwidth, and this implies a lower level of noise within the input amplification circuitry. Consequently there would appear to be merit in choosing a pulse shape that keeps noise to a minimum (low bandwidth) yet remains well defined at the decision instants.

Normally the equalization process is performed in two stages. A *preequalizer* generates a suitable pulse shape, and provides initial compensation for the transmission medium. Further pulse shaping and final compensation for the medium occurs at the *postequalizer,* situated at the distant end of the transmission path. This approach is usually simpler to analyze compared to designing an equalizer that performs the required functions in a single step.

## 8.2.5 Line Build Out (LBO)

We have seen in the previous section that the equalizer must have a frequency-attenuation transfer function which is the inverse of that for the cable. This will be difficult to arrange if cables of different lengths are fitted between the repeater sections, since it implies a transfer function of variable slope.

Eq. (8.5) makes it clear that the equalization of varying cable lengths would be difficult to perform within a single unit. Consequently a two-stage approach is to be recommended. The first stage, the *line build out* (LBO) unit, functions as a variable length cable simulator, duplicating the frequency-attenuation/phase transfer characteristic for a specified length of cable. The second stage contains an equalizer with a transfer function related to a predetermined reference length of cable. Thus the transfer functions of the actual cable correspond to that of the LBO, plus equalizer.

The LBO units within early equipments were made in a range of fixed values. These were then selected appropriately during the installation of each repeater.

---

[1] Values refer to a high-quality symmetric-pair cable and have been obtained from: J. Doemer and R. Pospischil "Communications Engineering," *Siemens Review,* special issue, vol. XL1, 1974, p. 281.

This was a time-consuming and expensive process. Furthermore, if an incorrect value of LBO was selected by mistake, the result would be a difficult to identify, degraded performance of the repeater.

Current repeater designs employ *automatic line build out* (ALBO) units that can compensate for a wide range of attenuations. These devices quantize the level of the received pulses using a peak detector, and switch different passive elements, plus various amplification stages, in and out of circuit as required. In this way the transfer function of the circuit can be modified and equated to different cable lengths automatically.

The practical realization of an ALBO circuit is much more difficult than the above description would suggest, especially if a large attenuation range is required. For example, in the United Kingdom the ALBO must compensate for a 37-dB attenuation range (first-order line equipment), and this is extremely difficult to achieve. Furthermore, special attention must be paid to the effects of crosstalk when such high amplifier gains are involved.

Unless special precautions are taken the ALBO will move to its most sensitive range when the line signal fails. In this case if the crosstalk from other digital signals has any significant level, the ALBO will amplify them, and pass them as a valid input to the regenerator. This chain of events is very serious since the terminal alarm equipment will not be aware that a fault has occurred.

The problems described above closely correspond to those encountered in the development of analog FDM transmission systems. However, in the digital case, high linearity assumes less importance compared to good transient response.

## 8.3  SYMMETRIC-PAIR AND COAXIAL CABLES

We have already seen in Chap. 2 that the length of a transmission path may range from one or two to several thousand kilometers. Obviously it is not practical to satisfy this requirement by making and laying the required cable in one continuous length!

In most countries (e.g., the United States, United Kingdom, Italy, and South Africa) standard practice is to lay cables inside underground ducts. In this case the maximum primary cable length will be determined by the practical limits that may be drawn into the ducts via the special access points provided. Typical lengths lie in the range 100 to 400 m.

Some countries mount their telecommunication cables and repeaters above ground (e.g., Brazil); while others bury their cables without the use of ducts (e.g., Belgium). In this case primary lengths will be dependent on the practical limitation of cable transportation to the cabling site.

### 8.3.1  Cable Construction

Traditionally, symmetric-pair audio quality cable used a copper conductor, a dry-paper insulator, and a lead protective sheathing. Over the years polyethylene has replaced paper as an insulation material, while aluminum conductors and plastic sheathing have been introduced for reasons of cost. The individual wires were arranged as twisted pairs, or as groups of four wires (star quads) twisted together. The twisted assemblies were then placed in concentric layers to complete the cable.

Experience has shown that the rate of twist must be randomly varied along the length of the cable, or a low-frequency component may be induced into the signal path.  Crosstalk between individual pairs may be minimized by varying the packing distance between the pairs.

Symmetric-pair cables are preferred to coaxial cable in the lower PCM hierarchy levels (1.5 to 8 Mbit/s) as they are cheaper to manufacture and increase the efficiency of power feeding for the same conductor cross section.  A major disadvantage, however, with symmetric-pair systems is that the crosstalk between individual pairs increases at high transmission frequencies.  Coaxial cables offer significant advantages in this respect, and are arguably preferable, taking the economics into account, at transmission rates above 10 Mbit/s.

In some countries, notably the U.S., Belgium, and Germany, there is already in existence a widespread network of high-quality pair cable.  These were originally used in conjunction with medium capacity FDM systems at frequencies up to about 250 kHz, equivalent to 60 telephony channels.  The pairs in these cables are arranged in groups (e.g., five pairs per group), and each group is shielded to minimize the effect of crosstalk.

Recently several telephone administrations have considered reequipping these high-quality cables with digital equipments.  Trials have been made at 6.3 Mbit/s in the U.S.,[2] 8.448 Mbit/s in Germany[3] and Belgium, and 17 Mbit/s in Belgium.  The results of such tests have shown that high-bit-rate transmission is possible using symmetric-pair cable.  However, many observers have indicated that coaxial cables are to be preferred at these frequencies, since repeater designs can be considerably simplified.

The coaxial cables that are employed on trunk routes within the telephone network use a semiair dielectric between the inner and outer (copper) conductors.  The inner conductor is supported by polyethylene discs at intervals of just over 3 centimeters (cm).  The outer conductor is formed by thin copper tape supported by a secondary layer of steel tape.  A third layer of paper tape insulates adjacent coaxial tubes from each other.

Different sizes of coaxial cable have been developed for analog FDM transmission; the most common size having an inner conductor diameter of 2.64 millimeters (mm), and an outer diameter of 9.5 mm.  A special cable has also been developed in Italy for digital transmission at 34 Mbit/s.  This cable is commonly known as *microcoax* and has an outer diameter of 2.8 mm.  The various types of coaxial cable that are in common usage today are listed in Table 8-1 together with their characteristic size and other relevant data.

## 8.3.2  Impedance Consideration

The impedance of the coaxial cables listed in Table 8-1 is approximately 75 ohms ($\Omega$), while symmetric-pair cables typically exhibit an impedance in the range 100 to 120 $\Omega$.  The following expression provides a close approximation for the frequency dependence of the line impedance parameter $Z$:

$$Z = Z_\infty + \frac{A}{F}(1 - j) \qquad \Omega \tag{8.7}$$

[2] D. E. Setzer, "Low Capacitance Cable for 6.3 Mbit/s Transmission Systems," *Proceedings of the Institute of Electrical and Electronics Engineers 1972 International Communications Conference,* Institute of Electrical and Electronics Engineers, Philadelphia, 1972.

[3] J. Doemer and R. Pospischil, "Communications Engineering," p. 280.

**Table 8-1  Coaxial Cable Data**

| Type of cable | Conductor diameter, mm | Attenuation at 1 MHz (10°C), dB/km | Possible usage | Nominal value S/N ratio for repeater, dB‡ |
|---|---|---|---|---|
| Microcoax | 0.65/2.8 | 9.5 | 8.448 Mbit/s; MS43 line code; 4 km repeater spacing* | 44.5 |
|  |  |  | 34.368 Mbit/s; MS43 line code; 2 km repeater spacing† | 37.9 |
|  |  |  | 139.264 Mbit/s; MS43 line code; 1 km repeater spacing |  |
| Minicoax | 1.2/4.4 | 5.3 | 34.368 Mbit/s; MS43 line code; 4 km repeater spacing | 35.7 |
|  |  |  | 139.264 Mbit/s; MS43 line code; 2 km repeater spacing | 29.0 |
|  |  |  | 565.148 Mbit/s; MS43 line code; 1 km repeater spacing | 22.3 |
| Standard coax | 2.6/9.5 | 2.35 | 139.264 Mbit/s; MS43 line code; 4–65 km repeater spacing | 22.5 |
|  |  |  | 565.148 Mbit/s; MS43 line code; 1–55 km repeater spacing | 48.8 |

\* Field trials have been made.

† In service.

‡ These values are given as an example only, and may be insufficient in some cases.

SOURCE: From L. Bellato, A. Tavella, G. Vannucchi, *New Digital Systems over Physical Lines*, Teletrra SPA., Milano, Italy.

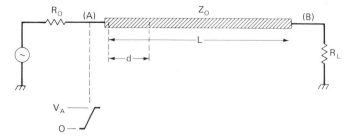

**Fig. 8-8**   Impedance matching the transmission medium.

where $Z_\infty$ = real impedance
$A$ = constant

Typical value[4] for microcoax = 71.8 $\Omega$
Typical value[4] for microcoax = 3.06 $\Omega$ at 1 MHz

In theory the line impedance value must be closely matched by the input and output impedances of the repeater. This ensures that the transmitted signal is launched efficiently into the cable, and that no reflections occur at the receiving end of the cable. It is interesting to analyze this statement more exactly.

Let us consider the setup in Fig. 8-8, where $R_0$ represents the output source impedance of repeater $A$, and the impedance of the transmission line equals $Z_0$.

Now according to classical theory, the pulse rise time at $A$ will be unaffected by the transmission line capacitance. Consequently the transmitted pulse will remain undistorted, apart from "skin effect" and dielectric losses, until it reaches the load $R_L$. The voltage waveform $V(x,t)$ going down the line may be represented as a function of distance and time by the equation:

$$V(x,t) = V_A(t)\ U(t - Dt_{pd}) \qquad \text{for } t < T_D \tag{8.8}$$

where $V_A$ = voltage at point $A$
$d$ = distance to an arbitrary point on the line
$t_{pd}$ = propagation delay of the line, in nanoseconds/unit length (ns/unit length)
$T_D$ = total propagation delay of the line, ns.
$U(t)$ = a unit step function occurring at $t = 0$

When the pulse of voltage $V_I$ reaches point $B$ it will be partially reflected if $R_L$ and $Z_0$ are unequal. Suppose that the voltage of the reflected pulse equals $V_R$, then the reflection coefficient at the load $\rho_L$ can be derived by using Ohm's law as follows:

$$\text{The voltage at the load } V_L = V_I + V_R \tag{8.9}$$
$$V_L = (I_I + I_R)\ R_L \tag{8.10}$$

Now,

$$I_I = \frac{V_I}{Z_0} \tag{8.11}$$

$$I_R = -\frac{V_R}{Z_0} \tag{8.12}$$

The negative sign indicates a pulse traveling towards the source.

[4] Ibid.

By substitution of Eqs. (8.11) and (8.12) into (8.9) and (8.10):

$$V_L = V_I + V_R = \left(\frac{V_I - V_R}{Z_0}\right) R_L \tag{8.13}$$

By definition

$$\rho_L = \frac{\text{reflected voltage}}{\text{incident voltage}} = \frac{V_R}{V_I} \tag{8.14}$$

By rearranging terms in Eq. (8.13) we get:

$$\rho_L = \frac{R_L - Z_0}{R_L + Z_0} \tag{8.15}$$

Using similar arguments we may obtain the following expression for the reflection coefficient at the source $\rho_S$:

$$\rho_S = \frac{R_0 - Z_0}{R_0 + Z_0} \tag{8.16}$$

The line voltage at any instant in time may be derived by summing the incident voltage $V_I$ together with the contributions from various orders of reflection ($B$ to $A$, $A$ to $B$, $B$ to $A$, etc.). In this case:

$$\begin{aligned}
V(x,t) = V_A(t)[&U(t - Dt_{pd}) + \rho_L U(t - t_{pd}(2L - D)) \\
&+ \rho_L \rho_S U(t - t_{pd}(2L + D)) + \rho_L \rho_S U(t - t_{pd}(4L - D)) \\
&+ \rho_L 2 \rho_{S2} U(t - t_{pd}(4L + D)) + \cdots]
\end{aligned} \tag{8.17}$$

In practical systems employing repeater separations of order 2 km there will be no problem due to multiple reflections, since these will be significantly attenuated by the cable. This will not necessarily be true for shorter connections, however, such as those used between equipments within the same terminal buildings. In this case it is good practice to include damping components within the repeater output circuitry that are arranged to attenuate second- and higher-order reflections.

The shape of the pulse received at $B$ can be considerably altered due to the effect of line reflections. This complicates the decision circuitry and can only be avoided by good impedance matching at the receiver in the first place.

### 8.3.3  Practical Considerations

The practical realization of matching the source and load impedances to that of the transmission medium is somewhat more difficult than the above theory would suggest. A typical circuit implementation is shown by way of example in Fig. 8-9.

Almost perfect matching is obtained at the receiver (load) provided that a mark is being transmitted. (A *mark* is the name given to the on state when a pulse is present.) However, a mismatch is permitted during the time interval between marks.

The situation at the transmitter (source) is somewhat different, since here we are concerned with transmission efficiency. A very large mismatch can be permitted if sufficient transmitter power is available.

It is instructive to examine the circuitry shown in Fig. 8-9, in more detail.

**8.3.3.1  Source.** The source impedance $Z_S$ is given by

$$Z_S = \frac{R1}{N^2} \tag{8.18}$$

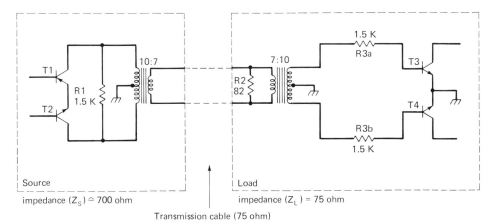

**Fig. 8-9** A typical implementation of source-load impedance matching. Overvoltage and other protective circuitry is omitted.

where $N$ = turns ratio of the transformer

Therefore, by substitution of the appropriate values:

$$Z_S = \frac{1.500}{(1.43)^2} \simeq 700 \ \Omega$$

In this case the source impedance is approximately 10 times that of the transmission medium! The resultant inefficiency is usually acceptable provided a large power reserve is available, and a simple circuitry implementation is achieved.

**8.3.3.2  Load.**  The load circuitry will exhibit two impedance values, dependent on whether a mark is present or not. If there is no mark the load impedance will be equivalent in value to $R2$, that is 82 $\Omega$. In this case both transmission $T1$ and $T2$ will be off.

If a mark is present, then dependent on its polarity, either transistor $T3$ (positive mark) or $T4$ (negative mark) will be driven into saturation and switch on. This alters the load impedance somewhat by putting $T3$ and the transformer in parallel with $T2$. In this case the load impedance $Z_L$ is given by

$$Z_L = \frac{R2(R3/N^2)}{R2 + (R3/N^2)} \tag{8.19}$$

$$Z_L = \frac{82 \times (1.500)/(1.43)^2}{82 + (1.500)/(1.43)^2} \simeq 74 \ \Omega$$

The reflection coefficient $\rho$ may be determined by substitution into Eq. (8.15). Therefore,

$$\rho = \frac{75 - 74}{75 + 74} = 0.01 \tag{8.20}$$

Normally it is necessary to calculate the power of the reflected signal in order to ensure that its effect can be ignored. This quantity is commonly referred to as the *return loss P*, and may be derived from the following expression:

$$P = 10 \log_{10} \left( \frac{1}{1 - |\rho|^2} \right) \qquad (8.21)$$

By substitution of Eq. (8.20) into (8.21) we get:

$$P = 10 \log_{10} \left( \frac{1}{1 - |0.01|^2} \right) \simeq 4 \times 10^{-4} \text{ dB} \qquad (8.22)$$

The return loss calculated in this example is very small. Consequently the effect of multiple reflections, or received signal degradation need not be investigated further.

Transmission line theory is described in several standard texts, and it is therefore inappropriate to dwell on this subject here. From a practical point of view the ITT publication *Reference Data for Radio Engineers* and the Motorola publication *MECL System Design Handbook* are to be recommended.[5]

## 8.4   ERROR RATE

The quality of communication within a digital transmission system may be conveniently expressed in terms of the *error rate* or the *error probability* $P_e$. However, a single numerical value for $P_e$ will normally be insufficient in itself to define the operating standards of a telephone network. The following values need to be separately identified:

1. DESIGN ERROR RATE[6]

    This is a measure of the performance of an *item* when operating in a specified electrical/physical environment. The features of this environment (temperature, power supply, noise, cable characteristics, electrical interference, etc.) may independently assume values confined to some range. The design error rate is used as a design objective for both the equipment and system design.

2. MEAN ERROR PERFORMANCE[6]

    This is a measure of the expected average performance of the equipment when in normal service.

3. BURST ERRORS

    The design and mean error rate values are calculated assuming purely random perturbations to the system. However, in a digital network periods of excessive error rate can also occur. The duration of these intervals may range from a few milliseconds in the case of impulse noise (e.g., car ignition interference) to a few hours in the case of abnormal functioning of the system (e.g., radio relay system fades).

    The effect of burst errors must be taken into account within the design of the system. For example it is usual to transmit certain data messages as packets which are retransmitted if corrupted by errors.

---

[5] Howard W. Sams Engineering Staff, *Reference Data for Radio Engineers,* 6th ed., Sams, Indianapolis, Ind., 1975; and W. R. Blood, Jr., *Motorola Emitter Coupled Logic (MECL) System Design Handbook,* Motorola Semiconductor Products, Phoenix, Ariz.

[6] CCITT, Special Study Group D, contribution no. 331, document no. 80, International Telecommunications Union, Geneva, Switzerland, July 1976.

4. EXCESSIVE ERROR RATE

This is used to determine when the performance of a piece of equipment, or a transmission path, has reached an unacceptable level, making maintenance necessary.

It is interesting to consider what a given error rate means in real terms. Subjective tests have shown that an error probability of order $10^{-5}$ is just discernible on low-level speech. Furthermore an error probability of $10^{-4}$ continues to provide a reasonable speech quality; however, at $10^{-3}$ the quality becomes unacceptable. These values have been used by the CCITT to provide advance notice of a gradual decay in the transmission quality. In this way maintenance action can be taken before the transmission path becomes unserviceable. Thus a *deferred maintenance alarm* is set at an error-rate threshold of $10^{-4}$, while *prompt maintenance* is required for error rates higher than $10^{-3}$.

Tests have shown that the transmission quality required for the communication of studio quality television and sound signals should be better than $10^{-5}$ depending on the encoding technique employed. This implies that standard telephony quality transmission paths cannot be employed for these signals, unless special precautions are taken. Studies are presently being made into the possibility of using error correction, or mitigation coding to overcome this potential problem.

### 8.4.1  Loss of Frame Alignment, and Error Rate

We have already seen in Chap. 6 that the effect of transmission errors must be taken into account when designing the mechanism for losing frame synchronism within a TDM multiplexer. Furthermore, we investigated in Chap. 7 the susceptibility of the justification control digits to such errors. Here we shall consider the error probability objectives for TDM multiplexers connected in tandem.

The average time interval $T_F$ between loss of frame alignment due to random digit error is given by

$$T_F = \frac{F}{P_n} \qquad (8.23)$$

where $F$ = time occupied by one frame period
$P_n$ = probability of $n$ consecutive frame alignment words each containing an error

$$P_n = (fP_e)^n \qquad (8.24)$$

where $f$ = number of bits in frame alignment word
$P_e$ = probability of any single digit being in error

By substitution of Eq. (8.24) into (8.23) we get:

$$T_F = \frac{F}{(fP_e)^n} \qquad (8.25)$$

This result has been used to plot Fig. 8-10 for different frame formats and alignment mechanisms. The CCITT recommended frame formats for the first-order (see Fig. 6-16), second-order, and third-order (Table 7-2) multiplexers are used as an example. In this case, after making the appropriate substitution Eq. (8.25) may be rewritten as:

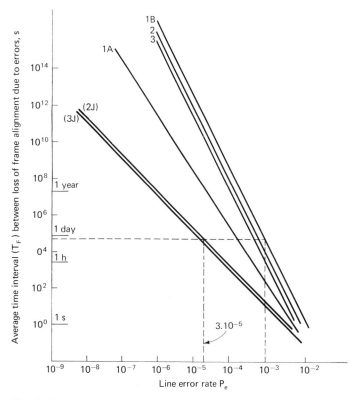

**Fig. 8-10** The impact of line error rate on justification detection. (Justification error: plots *1A, 1B, 2, 3*; justification alignment error: plots *2J, 3J*; see text for explanation.)

- Plot 1*A*, CCITT recommended first order G732 (*A* = 3):

$$T_F = \frac{512}{2.048 \times 10^{-6}} \frac{1}{(7P_e)^3}$$

- Plot 1*B*, CCITT recommended first order G732 (*n* = 4):

$$T_F = \frac{512}{2.048 \times 10^{-6}} \frac{1}{(7P_e)^4}$$

- Plot 2, CCITT recommended second order G742:

$$T_F = \frac{848}{8.448 \times 10^{-6}} \frac{1}{(10P_e)^4}$$

- Plot 3, CCITT recommended third order G751:

$$T_F = \frac{1536}{34.368 \times 10^{-6}} \frac{1}{(10P_e)^4}$$

A comparison between the four plots in Fig. 8-10 shows that frame alignment loss will occur at intervals of a few hours (plot 3) to approximately once a year (plot

1*b*) for an error rate of $10^{-3}$. Thus frame synchronism will be retained for a long time in spite of the transmission quality reaching an unexpectedly low level.

### 8.4.2 Loss of Justification, and Error Rate

It is interesting to compare the above results with the average time interval $T_J$ between misinterpretations of the justification code word, due to random digit errors. Let us consider a triplicated (3-digit) code word that is interpreted on the basis of a majority decision. In this case 2 digits must be in error for a misinterpretation to occur.

Let the probability of interpreting the justification code correctly $= P_J$. The probability of 1 digit being incorrect $P_I = P_e$. The probability of 1 digit being correct $P_C = 1 - P_e$. Therefore,

$P_J =$ probability of 2 out of 3 digits being correct + the probability of 3 digits being correct.

$$P_J = (P_C)_{\text{Digit } 1,2,3} \ (P_C)_{\text{Digit } 2,3,1} \ (P_I)_{\text{Digit } 3,2,1} + (P_C)^3 \tag{8.26}$$
$$P_J = 3(1 - P_e)(1 - P_e)P_e + (1 - P_e)^3 \tag{8.27}$$
$$P_J = (3P_e - 6P_e^2 + 3P_e^3) + (1 - 3P_e + 3P_e^2 - 3P_e^3) \tag{8.28}$$
$$P_J = 1 - 3P_e^2 + 2P_e^3 \tag{8.29}$$

The probability $P_M$ of misinterpreting the justification code $= 1 - P_J$. Therefore,

$$P_M = 3P_e^2 - 2P_e^3 \tag{8.30}$$

Normally $P_e$ will be very small in value, and consequently the cubic term in Eq. (8.30) can be neglected:

$$P_M = 3P_e^2 \tag{8.31}$$

Now, $T_J$ is given by the expression:

$$T_J = \frac{F}{P_M} \tag{8.32}$$

By substitution of Eq. (8.31) into (8.32) we get:

$$T_J = \frac{F}{3P_e^2} \tag{8.33}$$

This result has been used to plot curves 2*J* and 3*J* in Fig. 8-10, by making the following appropriate substitutions:

• Plot 2*J*, CCITT recommended second order G742:

$$T_J = \frac{848}{8.448 \times 10^{-6}} \frac{1}{3P_e^2}$$

• Plot 3*J*, CCITT recommended third order G751:

$$T_J = \frac{1536}{34.368 \times 10^{-6}} \frac{1}{3P_e^2}$$

It is interesting to note that the justification code will be misinterpreted on average every 10 or 1000 s (15 min) for error rates of $10^{-3}$ and $10^{-4}$, respectively. Each misinterpretation will result in a frame slip, and necessitate the resynchronization of the dependent multiplexers. Consequently a better error rate performance is required for the transmission of multiplexed signals that contain a 3-digit justification code. This point is illustrated in Fig. 8-10, where a uniform grade of service is seen to be obtained for minimum error rates of $10^{-3}$ (plot $1b$), and $3 \times 10^{-5}$ (plots $2J$ and $3J$).

The above analysis may be extended to cover the case of 5-digit majority-voted justification coding schemes. These are considerably more resistant to random digit errors.

### 8.4.3   Signal-to-Noise Ratio, and Error Rate (Binary)

Most noise signals are the result of numerous uncorrelated disturbances. Thermal noise, shot noise, and even the crosstalk introduced by neighboring transmission systems are our concern here. Disturbances of this type exhibit a truly random behavior, and it can be shown that the relative frequency of occurrence of individual noise amplitudes $x$ can be very closely approximated by a gaussian distribution $p(x)$ given by

$$p(x) = \frac{1}{\sqrt{2\pi}} \exp - (x^2/2\sigma^2) \tag{8.34}$$

where $\sigma = $ rms amplitude value of the noise signal $n(t)$.

The effect of noise in a digital transmission system is somewhat different to that observed in other communication methods. In binary transmission, which we shall consider here, we are interested in knowing whether a pulse is present or absent during a known time interval $T$. We are not interested in the shape of the received waveform, since this is already known. Essentially this means we need to analyze the probability of a significant noise contribution during the decision interval.

A decision threshold $d$ must be specified such that a pulse is considered present if its amplitude exceeds $d$. The message (binary signal) is illustrated in the time domain within Fig. 8-11, where the effect of noise is also shown.

The probability density function for the cases pulse absent and pulse present is illustrated within Figs. 8-12$a$ and $b$, respectively. It is clear that the probability of an error for the case pulse absent is given by the shaded area outlined by the probability curve (Fig. 8-12$a$). Thus, although a pulse may be absent there exists a finite probability that an erroneous interpretation will occur.

The converse case, when a pulse is present and an erroneous interpretation occurs, is illustrated in Fig. 8-12$b$. This type of error is called *false dismissal*.

The most usual pulse amplitude that occurs when the signal is present is $V$, although smaller and larger values are also possible due to the additive effect of gaussian noise when a signal is present. When the received pulse amplitude falls below $d$,

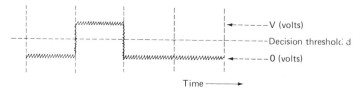

**Fig. 8-11**   Detection of binary pulses, by specifying a decision threshold.

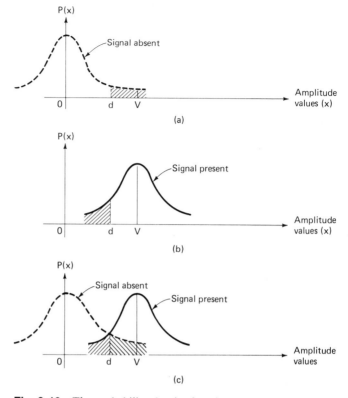

**Fig. 8-12**  The probability density function, associated with pulse detection.  Gaussian distributions are assumed for pulse present and absent conditions.  The decision threshold $d$ refers to Fig. 8-11.

an erroneous interpretation will clearly result, and this has a probability given by the shaded area in Fig. 8-12$b$.

The total error probability is given by the mean of the two shaded areas in Figs. 8-12$a$ and $b$.  It is clear from Fig. 8-12$c$ that the sum of the two areas reaches a minimum value when

$$d = \frac{V}{2} \tag{8.35}$$

This defines the optimum position for the decision threshold at half the pulse height.

In practical systems, however, other factors will influence the ideal threshold position. We may take note of the following in particular:

1. The transmitted pulse is considered to be a binary pulse of fixed amplitude $V$, duration $T$, within the above analysis.  In fact the treatment is generally valid for pulses of any shape that are disturbed by gaussian noise alone.  However, we do not take into account here the effect of echo, intersymbol interference, or threshold decision errors.
2. We have already noted that digital signals are quantized in both amplitude and time.  The above analysis takes no account of the time variable; although

it is clear that judicious positioning of the sampling instant affects the probability of an error (see Chap. 10).

A more rigorous treatment taking the above factors into account yields an optimum threshold in the range 44 to 46 percent of the maximum pulse height for raised cosine pulses. (Theoretical analysis made by L. Korowajczuk, Telebrás, Research and Development Center S.P., Brazil.)

Let us continue our analysis, and consider Eq. (8.34) as valid. Furthermore, we shall assume that the shaded areas in Figs. 8-12a and b are equal. In this case the total error probability for random data is given by either of the areas. Consequently the total probability of error $P(e)$ may be obtained by integrating Eq. (8.34) over the region defined by the shaded area in Fig. 8-12a. Therefore,

$$P(e) = \int_d^\infty p(x)\, dx \tag{8.36}$$

By substitution of Eq. (8.34):

$$P(e) = \frac{1}{\sigma\sqrt{2\pi}} \int_d^\infty \exp - (x^2/2\sigma^2)\, dx \tag{8.37}$$

This integral is referred to as an *error function* (erfc), or *probability integral,* and cannot be evaluated in a closed form. If Eq. (8.37) is rewritten as an error function,[7] then

$$P(e) = \frac{1}{\sqrt{2\pi}} \int_{d/\sigma}^\infty \exp - (x^2/2)\, dx \tag{8.38}$$

$$P(e) = \text{erfc}\left(\frac{d}{\sigma}\right) \tag{8.39}$$

The error function [erfc $(x)$] may be approximated[8] to give:

$$\text{erfc}(z) \simeq \frac{1}{z\sqrt{2\pi}}\left(1 - \frac{1}{z^2}\right)\exp - (z^2/2) \qquad \text{for } z > 2 \tag{8.40}$$

This approximation is accurate to within about 10 percent for $x = 2$, and better than 1 percent for $x > 3$. By substitution of Eq. (8.39) into (8.40) we get:

$$P(e) = \frac{1}{\sqrt{2\pi}} \frac{\sigma}{d}\left[1 - \left(\frac{\sigma}{d}\right)^2\right]\exp - \frac{1}{2}\frac{d}{\sigma} \tag{8.41}$$

This expression relates the total error probability to the decision threshold $d$ and the rms amplitude of the noise signal $\sigma$. This must now be rewritten in terms of the signal-to-noise power $S/N$. The average signal power $S$ for a binary signal that has the levels $V$ volts (pulse present), and 0 volts (pulse absent), is given by

$$S = \frac{V^2}{2} \tag{8.42}$$

---

[7] Many authors use somewhat different definitions for the error function than that used here. These are essentially equivalent, with minor differences.

[8] J. M. Wozencraft, and I. M. Jacobs, *Principles of Communication Engineering,* Wiley, New York, p. 82.

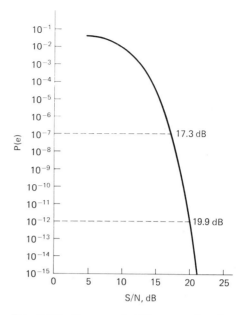

**Fig. 8-13** Error probability, as a function of signal-to-noise ratio.

By substituting Eq. (8.35) into (8.42) we get:

$$S = 2d^2 \qquad (8.43)$$

The average noise power $N$ is given by definition as

$$N = \sigma^2 \qquad (8.44)$$

Thus, the signal-to-noise power ratio $S/N$ is given by

$$S/N = 2(d/\sigma)^2 \qquad (8.45)$$

By rearranging terms we get:

$$\frac{\sigma}{d} = \sqrt{2(S/N)^{-1}} \qquad (8.46)$$

By substitution of Eq. (8.46) into (8.41) we get:

$$P(e) = \frac{1}{\sqrt{2\pi}} \left[ 2(S/N)^{-1} \right]^{1/2} \left[ 1 - 2(S/N)^{-1} \right] \exp{-\left( \frac{S/N}{4} \right)} \qquad (8.47)$$

This result is plotted in Fig. 8-13, where $S/N$ is measured in dBs.

A few details relating to this result are worthy of comment. First, the expression for $P(e)$ is only valid for single-polarity binary signals subjected to gaussian noise. Second, the approximation of the error function [Eq. (8.40)] is inaccurate for small $z$ values. The reader is advised, if more precision is required, to consult one of the standard error function tables.[9]

There exist in the literature many curves that are similar, but not identical to

[9] See for example: "Standard Mathematical Tables," Chemical Rubber Company, Cleveland, Ohio.

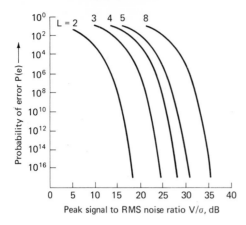

**Fig. 8-14** Error probability as a function of signal-to-noise ratio for different numbers of transmission levels $L$. (From Bell Telephone Laboratories Technical Staff, *Transmission Systems for Communications*, Bell Telephone Laboratories, Western Electric Co., 1971, p. 629. Copyright 1971, Bell Telephone Laboratories Inc., reprinted by permission.)

Fig. 8-13. This can be quite confusing, although the difference may be easily explained. Here we consider signal and noise power, not energy nor amplitude. Furthermore, many authors consider bipolar transmission pulses ($+V : 0 : -V$), which leads to a 3-dB disadvantage at all points on Fig. 8-13.

### 8.4.4  Signal-to-Noise Ratio, and Error Rate (Multilevel)

Most transmission systems employ ternary or high-order multilevel signals, rather than the on-off binary pulses envisaged above. Transmission signals to this type allow us to choose a suitable signal-to-noise power ratio for a given bandwidth, in accordance with the theoretical limits set by the Hartley-Shannon law (see Sec. 4.1.3).

Let us consider the transmission of pulses using an $L$ level transmission code. We shall assume that the levels occur with equal probability, and that they are equally spaced between $+V$, and $-V$ volts.

The separation between levels $\Delta v$ will be given by

$$\frac{2V}{L-1} \tag{8.48}$$

An error will occur in the uppermost and lowermost level each time the noise amplitude exceeds $\Delta v/2$. However, errors will occur at the intermediate levels for noise amplitudes of only $\Delta v4$. Consequently the total error probability for multilevel transmission $P_{ME}$ must be computed in two stages:

$$P_{ME} = \Sigma P_{OE} + P_{IE}$$

where $P_{OE}$ = probability of an error in the outermost levels
$P_{IE}$ = probability of an error in an intermediate level

In this case, using the same arguments developed above it can be shown that:[10]

[10] B. P. Lathi, *Communication Systems,* Wiley, New York, 1968.

$$P_{ME} = 2\,\frac{L-1}{L}\,\text{erfc}\,\frac{1}{L-1}\frac{V}{2\sigma} \tag{8.49}$$

The above relationship is plotted in Fig. 8-14, in terms of the peak signal-to-rms-noise ratio $V/\sigma$ measured in dBs.

It is interesting to note that in the case of binary transmission $L = 2$ there is a rapid increase in the error probability for a reduction in the signal-to-noise ratio below about 15 dB. This phenomenon is observed in any digital transmission system and is commonly referred to as the *error threshold,* or *cliff effect.*

In practical systems an error rate of order $10^{-10}$ per repeater section is required. This criterion demands a signal-to-noise power ratio at the receiver of order 16 to 17 dB.

### 8.4.5  Error Probability for *n* Repeater Sections

We have already seen that a reduction in the signal-to-noise ratio within a single repeater section in the case of regenerative transmission (see Sec. 8.2.3) will be critical for the total transmission path.

If each section has an error probability $P_e$, and there are $n$ sections in tandem, the total probability $P_S$ will be given by

$$P_S = \tfrac{1}{2} \times 1 - (1 - 2p_e)^n \tag{8.50}$$

This expression conveniently approximates for large $n$ values to

$$P_S = nP_e \tag{8.51}$$

Therefore, the error probability per section must be $n^{-1}$ of that for the total system.

The effect of regeneration is clearly seen in Fig. 8-15 where an analog scheme is

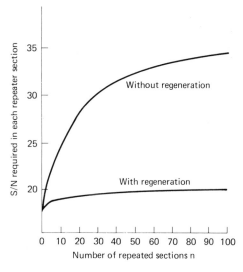

**Fig. 8-15**  The error probability of $n$ repeater sections. Signal-to-noise ratio assumes an error probability of $P(e) = 10^{-9}$. (From Bell Telephone Laboratories Technical Staff, *Transmission Systems for Communications,* Bell Telephone Laboratories, Western Electric Co., 1971, p. 218. Copyright 1971, Bell Telephone Laboratories Inc., reprinted by permission.)

shown by way of comparison. It is interesting to note that digital signals may be communicated using any of these three basic philosophies:

1. Regenerative repeatered sections (the most usual method).
2. Nonregenerative repeatered sections: Each repeater amplifies the transmission signal to its original level only. There is no clock extraction or regeneration performed. (This method is very unusual.)
3. Hybrid sections: Alternate sections are regenerative and non-regenerative, respectively. (This method has increasingly been used for very high-speed transmission.)

The hybrid scheme is particularly attractive for use at transmission bit rates of 500 Mbit/s and above, when media attenuation is particularly severe. This method has also been proposed for transmission via optical fiber at much lower bit rates.

### 8.5 THE EYE DIAGRAM

In the last section we related the error performance for the transmission of binary signals to the signal-to-noise power ratio. Normally, however, digital transmission systems employ pulse-shaping techniques in order to improve the transmission properties of the signal and to minimize intersymbol interference. Typically an information signal of the form illustrated in Fig. 8-16 will be transmitted, and this will normally exhibit a random behavior.

A favorite technique used by engineers during the evaluation of a digital system is to display the random transmission waveform on an oscilloscope triggered at the transmission bit rate. This has the effect of superimposing various transitions of the signal on top of each other, and produces a waveform as shown in Fig. 8-17. The resultant pattern is commonly referred to as an *eye diagram* for obvious reasons.

The effect of severe intersymbol interference, or noise disturbance is shown very

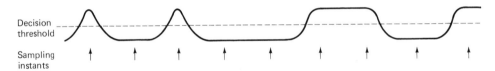

Decision threshold

Sampling instants

**Fig. 8-16** A typical random information signal.

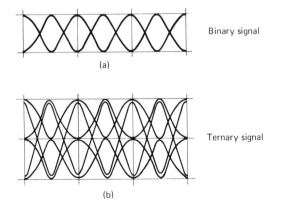

(a)

Binary signal

(b)

Ternary signal

**Fig. 8-17** The eye diagram.

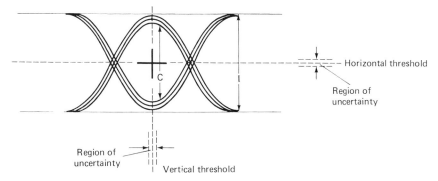

**Fig. 8-18**   The effect of the eye closure.

clearly by the eye diagram.   Such disturbances have the effect of closing the eye opening, and increase the probability of an error, as shown in Fig. 8-18.

The performance of the regeneration process may be evaluated by comparing the eye diagram against both the amplitude level decision threshold and the sampling instant.   This is analogous to centering cross-wires, here the level and time variables, against a target, the eye center.   This is illustrated in Fig. 8-18.

In practice, the decision threshold level will vary due to temperature effects, component tolerances, and the like.   Futhermore the relative level of the received waveform will be dependent on the matching achieved between the equalizer circuitry and the transmission medium.   These effects cause an uncertainty in the vertical plane (level).

Inaccuracies in the horizontal plane (timing) can also occur.   We shall discuss this possibility in Chap. 10, when we investigate the problem of pattern induced jitter.

The evaluation of digital transmission systems in the laboratory by measurement of the eye diagram opening should be approached with caution.   This technique can distinguish between a very poor regenerator (e.g., $P(e) = 10^{-3}$), and a good quality unit (e.g., $P(e) = 10^{-10}$).   However, this method cannot be used to distinguish between two high-quality regenerators, since in this case very slight, occasional eye closures will pass unnoticed on the oscilloscope.

It is frequently desirable to analyze different system and circuitry approaches during the development phase of a digital transmission equipment.   In this case considerable theoretical performance data can be obtained by computing the eye openings for a mathematical model of the system.   This involves analysis of the impulse response of each circuit element, and leads to fairly complex mathematics.

Evaluation of the theoretical eye, as opposed to that measured on an oscilloscope, can give very meaningful performance data.   For example, reliable error-rate information and the effect of component inaccuracies can be predicted for a worst-case design.

### 8.5.1   Eye Closure

We have already noted that eye closure can be attributed to several factors.   The most significant contribution in the vertical plane is normally caused by noise, as is illustrated in Fig. 8-18, where the ideal and closed eye openings are referred to as $I$ and $C$, respectively.

If the same error performance is required for the closed-eye case as the open case, the signal-to-noise ratio must be increased by

$$(S/N)_{\text{Increase, dB}} = 20 \log_{10} \frac{I}{C} \tag{8.52}$$

### 8.5.2 Multilevel Eye Diagrams

We have already noted that most digital transmission systems employ multilevel coding structures. Normally, ternary transmission ($L = 3$) is used, and this is illustrated within Fig. 8-17$b$.

Let us consider the more general case of multilevel transmission consisting of $L$ levels, equally spaced between 0 and $V$ volts. In this case $L - 1$ separate eyes will be formed and each will occupy an ideal vertical height $I$ of

$$I = \frac{V}{L-1} \tag{8.53}$$

It is useful to define the normalized eye degradation $D$ by the relationship

$$D = \frac{I-C}{V} \tag{8.54}$$

From Eqs. (8.53) and (8.54) we get:

$$D = \frac{I-C}{I(L-1)} \tag{8.55}$$

$$\frac{I-C}{I} = D(L-1) \tag{8.56}$$

Now, Eq. (8.52) may be rewritten as

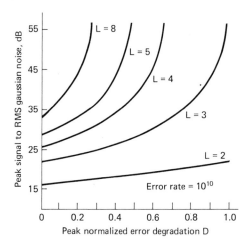

**Fig. 8-19** Signal-to-noise ratio as a function of the normalized eye degradation $D$. (From Bell Telephone Laboratories Technical Staff, *Transmission Systems for Communications,* Bell Telephone Laboratories, Western Electric Co., 1971, p. 633. Copyright 1971, Bell Telephone Laboratories Inc., reprinted by permission.)

$$(S/N)_{\text{Increase, dB}} = -20 \log_{10} \left[ 1 - \left( \frac{I - C}{I} \right) \right] \qquad (8.57)$$

By substitution of Eq. (8.56) into (8.57) we get:

$$(S/N)_{\text{Increase, dB}} = -20 \log_{10} \left[ 1 - D(L - 1) \right] \qquad (8.58)$$

This relationship between the total signal-to-noise power ratio and the normalized eye degradation $D$ has been plotted in Fig. 8-19 using Eq. (8.58). An error probability of $10^{-10}$, a typical value per repeater section, and gaussian noise disturbance has been assumed in plotting the graph.

It is important to once again state that the analysis for eye closure performed here takes account of gaussian noise disturbance only. The effect of intersymbol interference, echoes, jitter, and the practical tolerances of the implementation circuitry are ignored here. Readers who wish to pursue the problem of eye closure in further detail are recommended to consult one of the standard texts.[11]

## 8.6 CROSSTALK

Crosstalk is the most significant limitation within a symmetric-pair cable transmission system. It is useful to define two classes of crosstalk as follows:

1. NEAR-END CROSSTALK (NEXT)
   This is a function of the coupling between pairs, measured at the same end of the cable as the signal source.

2. FAR-END CROSSTALK (FEXT)
   This is a function of the coupling between pairs, measured at the remote end of the cable from the signal source.

The dominant crosstalk contribution for two-way transmission within the same cable sheath will always be NEXT. The actual value for NEXT can vary considerably between cables, although 68 dB may be regarded as a typical mean value for a range 50 to 85 dB.[12] In this case if the cable attenuation between two repeaters is, say, 40 dB, the NEXT interference will be on average at a level 28 dB below the received signal. However, in some cases, the NEXT interference will be only 10 dB below the received signal, equivalent to an error probability of order $10^{-3}$ for the offending section.

It is usual to avoid NEXT by isolating the two directions of transmission. This may be achieved by screening separate sections within the same cable sheath, or by using separate cables. If this is done, FEXT will be dominant.

A typical mean value for FEXT may be taken as 54 dB, for a range of values 38 to 70 dB.[12] Consequently the FEXT level is typically 14 dB below the NEXT value for the type of cable considerable here. It should be noted that it is important to distinguish between the signal-to-FEXT-noise ratio (e.g., 54 dB), and the absolute value of the FEXT attenuation. The absolute FEXT attenuation takes into account

---

[11] R. W. Lucky, J. Salz, and E. J. Weldon, Jr., *Principles of Data Communication,* McGraw-Hill, New York, 1968, pp. 65–67, 87; and J. S. Mayo, "Bipolar Repeater for Pulse Code Modulation Signals," *Bell System Technical Journal,* vol. 41, 1962, pp. 25–97.

[12] Values refer to British Post Office 74-pair cable; see G. H. Bennett, *Pulse Code Modulation and Digital Transmission,* Marconi Instruments, Saint Albans, Hertfordshire, England, 1976.

the cable attenuation also (e.g., 40 dB) and has a very low level (e.g., 54 + 40 = 94 dB).

Crosstalk increases at the rate of 6 dB per octave for FEXT, and 4.5 dB per octave for NEXT, thus the problem becomes more serious at high frequencies. It is this fact that has limited the use of unscreened pair cables to the lower-bit-rate transmission system such as the Bell T1, and the European 2 Mbit/s systems. At higher transmission frequencies coaxial cables, which exhibit negligible crosstalk, or unit-shielded pair cables, which exhibit some FEXT, must be used.[13]

The crosstalk introduced within a given pair, of a multipair cable, is due to the mutual coupling between many other energized pairs. Consequently the crosstalk will be dependent on the number of interferers and the efficiency of the coupling. The coupling efficiency will be dependent on the cable fabrication process and strict quality assurance must be carried out to keep this parameter to a minimum.

One of the attractions of introducing digital transmission within the telephone network has historically been the possibility of upgrading audio quality pair cable to carry multiplexed PCM signals (see Chap. 1). However, it should be noted that this upgrading cannot be performed indiscriminately or severe crosstalk problems could result. Pairs should be selected within the cable sheath according to minimum crosstalk design rules. Furthermore each selected pair must be tested for NEXT, FEXT, and overall attenuation prior to being approved for the transmission of digital signals. It is interesting to note that in certain countries, where labor costs are high (e.g., Canada), it has already become more economic to introduce PCM on *new* cables and avoid the selection and testing phase.

## 8.7 PRACTICAL CONSIDERATIONS

We shall now consider the additional circuits that are required for the practical realization of a digital transmission system.

### 8.7.1 Power Feeding

Typical transmission paths are alongside roads, across fields, and may be above or below ground. Consequently it is reasonable to assume that a power supply will not be locally available at each repeater site. This problem is usually overcome by passing a dc supply over the same cable pair used for conveying the digital signals. In coaxial systems it is usual to include within the same cable sheath containing the coaxial tubes, a special set of pairs for power feeding, and supervision (Sec. 8.7.3).

Power-feeding schemes must be carefully engineered to ensure the safety of maintenance personnel that may be working on the line system. By way of example the following limits have been specified by the British Post Office (BPO) for analog coaxial line systems:

- The current through a single 2000-$\Omega$ resistor (the assumed nominal resistance of a man), connected between any two current-carrying conductors, or from any conductor to earth, should never exceed 50 nanoamperes (nA) for more than 250 ns.
- The current through a 2000-$\Omega$ resistor connected in series with any current-

---

[13] L. Bellato, A. Tavella, G. Vannucchi, "New Digital Systems over Physical Lines," Telettra SPA., Milano, Italy.

carrying conductor should never exceed 50 milliamperes (mA) for more than 250 ns.

The BPO, therefore, employs a 50-mA constant current supply, with additional safety circuitry included within the design of the power-feeding unit. A maximum voltage of 75 V—0—75 V is used, offering 150 V between the pairs but restricting the voltage above earth potential to 75 V.

The power-feeding circuitry within each repeater normally consists of a chain of Zener diodes placed between the repeater input and output ports. The required voltages are selected at appropriate points within the diode chain. In practice the design of such circuits can be quite difficult for the following reasons:

1. Variations in the repeater transmission circuitry loading may alter the voltage levels supplied by the power-feeding unit quite considerably.
2. A feedback loop may easily be formed between the repeater input and output terminals. As a result, oscillation could easily occur in cases where the repeater has a high gain. This problem is particularly serious in coaxial cable systems where it is usual to include power-separation filters to avoid the problem.[14]

### 8.7.2 Overvoltage Protection

Todays repeater and line terminal equipments make extensive use of sophisticated semiconductor devices. Many of these have a very small geometric area, containing as many as 1500 transistors on a single integrated circuit chip. As a consequence of their reduced size, their ability to dissipate power is also reduced and some form of protection against occasional transient voltages must be provided. This is commonly referred to as *overvoltage,* or *lightning protection.*

Both exposed and buried telecommunications cables are sensitive to transients caused by:

1. Atmospheric discharge (lightning) resulting in (1) a direct hit, or (2) a hit in the vicinity of the cable causing magnetic or electrostatic induction within the cable.
2. Induction caused by high-voltage electric power lines.

A direct hit by lightning normally destructs both the cable and any associated electronic equipment, since it is extremely difficult to provide protection against this case. On the other hand, a hit in the vicinity of the cable will not normally cause cable damage; however, damage due to induced transient voltages could nevertheless result.

The magnitude of these induced voltages is dependent on the following parameters:

1. The magnitude of the atmospheric discharge current
2. The distance between the cable and the discharge point
3. The local soil resistivity
4. The construction of the cable (screening, etc.)
5. The type of cable ducting used, if any

Typically the discharge current ranges from a few kiloamps to a peak of 400 kiloamps (kA); a value that normally occurs only in tropical regions due to the greater height

---

[14] B. Hall and J. A. Jackson, "A New Power Feeding System for Coaxial Line Systems," *Post Office Electrical Engineers' Journal,* Jan. 1974.

from the bottom of the thundercloud. The resultant ground current is related to the soil resistivity, which can take a range of values from $2 \times 10^2$ to $2.5 \times 10^6$ $\Omega$ per cm depending on the content and material of the soil.

Now, if the upper soil layer happens to exhibit a high resistivity the discharge currents will follow any low-impedance path, such as a cable, to the lightning stroke point. This is the most common overvoltage induction mechanism due to lightning.

It is clear from the above that the requirements for lightning protection will vary from country to country. In South Africa for example, where soil resistivity is low, and lightning strikes are both severe and common, the problem has been solved by screening telecommunication cables in "earthed" metal ducts. In the U.S. several investigations[15] into the problem have resulted in the following worst-case[16] specification for the transient voltage that all repeater equipments must withstand:

Peak voltage:   800 V
Rise time:          10 microseconds ($\mu$s)
Fall time:          1000 $\mu$s

Until a few years ago it was believed that surges induced by lightning were the main, or perhaps the only cause of the line equipment failure. It is now realized that inductions from faulty neighboring high-voltage power lines can cause currents that are almost more important. Power-line faults can typically last up to 0.5 s, the response time of the power circuit breakers. Studies into this type of fault have indicated that the maximum transient voltages are considerably lower than that experienced for lightning discharge. However, the duration of the fault is longer (0.5 s), and currents of up to 20 amperes (A) are involved.

The above analysis would seem to suggest that two types of protection circuitry are required; these may be usefully defined as the primary and secondary protection circuits. The *primary protection circuit* reduces the effect of high-voltage transients (lightning discharges), and typically utilizes spark-gap gas protectors. The *secondary protection circuit* reduces the remaining transient to a level that is acceptable for semiconductor components, typically a Zener diode, a Trans Zorb,[17] or similar such device is used.

A typical protection circuit is shown, by way of example, in Fig. 8-20.

Primary protection is provided in this example, using a combination of three-, and two-electrode gas protectors. Secondary protection is provided by the resistors $R1$ and 2 and Zener diode $Z2$, which must be specifically selected to withstand large power dissipations (75 W typically). Zener diode $Z1$ is used for power feeding, and should also be capable of withstanding sudden power surges (5 W typically).

The additional capacitance introduced by the spark gaps is of order 1.5 to 2.5 picofarads (pF), and can be ignored in the case of low-bit-rate systems. However, at higher bit rates ($> 30$ Mbit/s) it must be accounted for within the design of the equalizer and the ALBO units. Consequently the development of a high-capacity line system often begins at the overvoltage protection circuit rather than the equalizer or ALBO. The converse is true for low-bit-rate systems.

[15] D. W. Brodle and P. A. Gresh, "Lightning Surges in Paired Telephone Cable," *Bell System Technical Journal,* Mar. 1961.

[16] A more severe test waveform has been specified by the CCITT; see Draft Recommendation K17 (Question 21/V).

[17] A trademark of General Semiconductor Industries: see O. M. Clark, "Transient Suppression Seminar: Supplementary Notes, 1976," General Semiconductor Industries, 2001 West Tenth Place, Tempe, AZ.

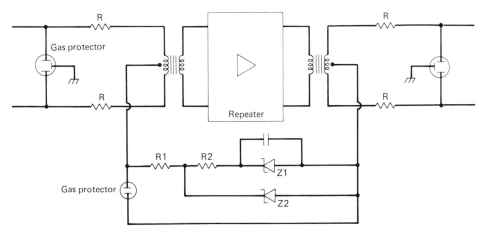

**Fig. 8-20** A typical example of overvoltage protection applied to a repeater.

### 8.7.3 Line Supervision and Fault Location

When a fault occurs within a repeater or transmission cable its location must be determined as rapidly as possible. Furthermore it is desirable from the maintenance point of view to have available continuously updated error-rate information about each transmission path. These features permit a standard grade of service to be maintained, and ensure that the economic loss due to breakdowns is kept to a minimum. The essential requirements for a supervisory system may be summarized as:

1. Transmission bit error-rate performance monitoring
2. Location of faults to within one repeater section
3. Raising a maintenance alarm when the transmission quality becomes unacceptable

It is useful to consider the two processes, detection of a failure, and the communication of the failure to the terminal equipment, separately. Indeed, in practice, these operations are completely separate, although the term *supervision circuitry* is applied to the two functions jointly.

The communication of a transmission failure or a degradation in the transmission quality may be conveniently transmitted to the terminal using specially designated *supervision pairs*. Some schemes automatically transmit supervisory data whenever a malfunction is detected, while others only transmit data when interrogated by the terminal. However, in both cases it is necessary to allocate a sequence of digits that uniquely identifies, that is, addresses each repeater. It should be noted that this address may be quite long since it is usually necessary to identify the faulty repeater within an access hole containing several parallel systems.

Supervision systems are commonly classified as being *in service* or *out of service*. The former system permits the checking of equipment while it is conveying normal information signals. The latter refers to those systems that require the link to be taken out of service in order that special test signals may be transmitted.

**8.7.3.1 In-Service Supervision.** The regular predictable form of a digital signal offers many advantages compared to other transmission methods for in-service supervision. We have already seen that the transmission of digital signals via pair, or coaxial cables requires pulses that process a well-specified shape, amplitude, level,

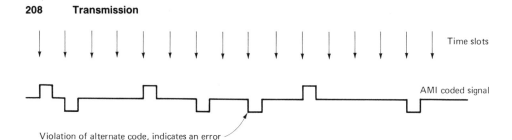

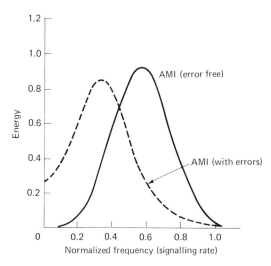

**Fig. 8-21**   Error recognition by detecting code violations.

**Fig. 8-22**   The modified energy density spectrum due to occasional errors.   The plots assume the transmission of random data.

and energy spectrum.   Furthermore we have noted that the dc level and energy spectrum may be controlled by selecting a suitable line code; while the amplitude and pulse shape parameters are governed by the adopted equalization characteristics (e.g., raised cosine) of the line.   The introduction of noise, and consequent transmission errors within the system will result in:

1. A disagreement, or violation, between the regenerated line signal, and the specified coding structure.   For example, if a bipolar (AMI) line-coding scheme is adopted, any single error will cause two consecutive marks to have the same polarity, as shown in Fig. 8-21.

2. A change in the mean dc level of the line signal due to an unequal number of marks with opposite polarity.

3. A change in the energy spectrum of the line signal.   This typically results, for *random*[18] data, in a significant increase in the low-frequency component of the signal, as shown in Fig. 8-22.

4. Degradations in the amplitude and pulse shape of the line signal.   This will normally result in errors, and leads to (1), (2), and (3) above.

[18] If the random nature of the transmission signal cannot be guaranteed, a specially generated test pattern must be used. This can only be implemented if the equipment is taken out of service.

The changes in the characteristics of the transmitted signal listed above may be readily detected and interpreted as indicative of faulty operation.

Disturbances of type (1) may be conveniently detected using simple logic circuitry, while analog detection methods may be employed for types (2) and (3). For example, a simple peak detector may be used to indicate when the line signal has moved outside its normal region of operation [type (2)]; or alternatively the line signal may be tested with a tuned circuit that is arranged to detect a significant low-frequency component [type (3)]. Generally speaking, analog detection methods are to be preferred since they tend to consume less power.

**8.7.3.2 Out-of-Service Supervision.** The most common method of error checking within a digital line system is based on transmitting a known pseudorandom

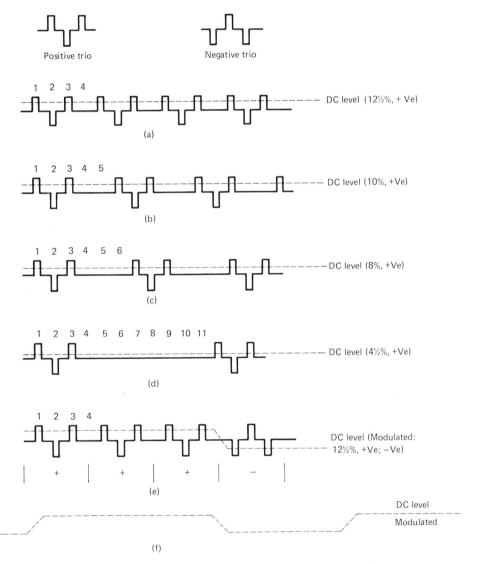

**Fig. 8-23** Supervision using "trio" test patterns: (a) ¼ unmodulated; (b) ⅕ unmodulated; (c) ⅙ unmodulated; (d) ¹⁄₁₁ unmodulated; (e) ¼ modulated; (f) resultant modulated dc level.

sequence, and comparing it against an identical reference sequence at points along the transmission path. This concept cannot easily be used for locating a faulty repeater section, and is normally only employed for checking the complete transmission path or during the commissioning phase.

An alternative approach to out-of-service monitoring has been used by the BPO within their 24-channel PCM system. This technique has been adopted in many countries in spite of certain drawbacks, due to the simplicity and low cost of implementation.

A special test pattern must first be generated with a pulse amplitude, width, and timing (clock) equivalent to the normal transmission signal. However, the code structure is deliberately violated in a well-defined manner, and causes gradual shifts in the mean dc level of the signal. More precisely triplet or trio pulse sequences are generated and modulated, as shown in Fig. 8-23.

The maintenance engineer may control the trio pattern in two ways:

1. The pattern density may be continuously varied over the range $\frac{1}{11}$ to $\frac{1}{4}$.
2. The mean polarity of the trio pattern may be inverted at some predetermined rate. In the case of the BPO, 18 preset frequencies are used occupying the range 1.1 to 3.2 kHz.

Each repeater in the transmission path contains a filter with a different frequency chosen from the 18 referred to above. The filter output is first amplified, and then connected to a common supervisory pair.

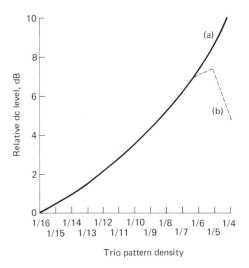

**Fig. 8-24** Typical relationship between the relative dc level and the trio pattern density: (a) curve for normal repeater operation; (b) curve for discontinuity indicates faulty repeater operation. (From A. N. Ramsden, "PCM Testing Technique," *Marconi Instrumentation*, vol. 12, no. 4, 1969.

The test is performed as follows:

1. One of the 18 modulation frequencies is selected.
2. The trio density is gradually increased from $\frac{1}{11}$ to $\frac{1}{4}$, while the dc level at the supervisory pair is monitored. Ideally the relative dc level will gradually increase with increasing trio-pattern density, as shown in Fig. 8-24.[19]

This method does have several shortcomings however. Most importantly, only the nearest faulty repeater within the transmission path can be exercised; subsequent repeaters can only be tested after the first fault has been corrected. Furthermore it is clear that the maximum number of repeaters that can be monitored by any terminal is 18. This limits the maximum number of repeaters to 36 for any given transmission path; a serious limitation in sparsely populated rural areas! Finally, some practical measurement problems have been observed on certain routes due to noise and crosstalk interference from other links.

[19] D. E. Waddington, "PCM Link Maintenance," *Marconi Instrumentation,* vol. 11, 1967, p. 2; and G. H. Bennett, "Testing Techniques for 24 Channel PCM Systems," *Post Office Electrical Engineers' Journal,* vol. 65, 1972, p. 182.

# 9

# Transmission Codes

In the previous chapter we noted that binary signals cannot normally be transmitted directly; they must first be coded into a form that overcomes the restrictions imposed by the transmission equipment.

We shall consider here the basic features of transmission coding, and how certain transmission parameters can be optimized by judicious code selection. The most important codes in use today are presented within the later sections of this chapter.

## 9.1 CODE PARAMETERS

Typically the coding process shapes a binary information signal at the transmitter site, to form a new signal that is more fit for transmission. At the receiver site the reverse process occurs, and the original binary information signal is retrieved. This simple analysis leads us to the following conclusions:

1. The coding scheme must be transparent to all binary information signals, since we have imposed no restrictions on the types of messages that may be transmitted.
2. The coded signal must be capable of being uniquely decoded to produce the original binary information signal.

We may also consider the requirements of the coded signal from the transmission point of view. These have already been identified within Chap. 8, and are summarized below:

3. There shall be no significant dc component.
4. A significant number of zero crossings shall be available for extraction of timing information (see Sec. 9.3).
5. The energy at low frequencies shall be small (see Sec. 9.3).
6. Invalid signal conditions should be detected easily, and used to indicate the transmission performance.

The coding process that satisfies the above requirements must not be allowed to significantly degrade the system performance on its own account. For example, single digit transmission errors will undoubtedly disturb the decoding process, and will result in multiple errors within the binary information signal. The multiplication of errors should be kept to a minimum.

The coding efficiency will also be important from the system point of view. Typically the information carrying capacity of the coded signal will be greater than the actual binary message information; the difference in the two quantities is used to accommodate the shaping of the coded signal. This implies that there must be an inherent inefficiency within the coding process, and it is clear that this should be kept to a minimum. We may summarize these considerations as follows:

  7. Low error multiplication.
  8. Good coding efficiency.

Finally we should note that it is possible to define a coding structure that sacrifices transmission efficiency for the facility of permitting the detection, or even the correction of occasional errors. There are many ways that this can be implemented, and several schemes have already been proposed in this book (e.g., majority voting of justification codes). We shall not pursue this topic further, however, since error-correction transmission coding has found limited applications only, and is the subject of extensive literature already. We may summarize this consideration as follows:

  9. Error-detection or correction capability.

We shall now consider the more important coding properties listed above in more detail.

## 9.2  CODING EFFICIENCY

The efficiency of a transmission code may be determined by comparing the information capacity, that is entropy, of the message signal and its coded equivalent.

Let us consider the passage of a quantity of information messages $a_1$, $a_2$, . . . , $a_n$, each having a probability of occurrence $p_1$, $p_2$, . . . , $p_n$. In this case it can be shown that[1] if a sequence of $N$ messages is transmitted, the total information $I$ communicated is given by

$$I = N \sum_{i=1}^{n} p_i \log_2 \frac{1}{p_i} \tag{9.1}$$

This may be expressed in terms of the information transmission rate $H$ given by

$$H = \sum_{i=1}^{n} p_i \log_2 \frac{1}{p_i} \qquad \text{bit/symbol} \tag{9.2}$$

If the message signal has a symbol rate of $S_m$ symbols per second, then the information capacity $C_m$ of this signal is given by

$$C_m = S_m \sum_{i=1}^{n} p_i \log_2 \frac{1}{p_i} \qquad \text{bit/s} \tag{9.3}$$

Similarly, the information capacity $C_c$ of the coded signal is given by

$$C_c = S_c \sum_{i=1}^{N} q_i \log_2 \frac{1}{q_2} \qquad \text{bit/s} \tag{9.4}$$

---

[1] C. E. Shannon, "A Mathematical Theory of Communication," *Bell System Technical Journal,* vol. 27, 1948, pp. 379–423, 623–656; and M. Schwartz, *Information Transmission, Modulation and Noise,* McGraw-Hill, New York, 1970.

where $S_c$ = the symbol rate, in symbols per second

$i$ = the probability of occurrence of the $i$th message symbol

The efficiency $\eta$ is given by definition as

$$\eta = \frac{C_m}{C_c} \tag{9.5}$$

By substitution of Eqs. (9.3) and (9.4) into (9.5) we obtain:

$$\eta = \frac{S_m \sum\limits_{i=1}^{n} p_i \log_2(1/p_i)}{S_c \sum\limits_{i=1}^{n} q_i \log_2(1/q_i)} \tag{9.6}$$

Let us suppose, for example, that a given coding scheme has the following characteristics:

Message signal     Binary ($n = 2$) transmission, at a rate of 2 MHz

                      Probability of a logical 1 = 0.4

                      Probability of a logical 0 = 0.6

Coded equivalent    Ternary ($N = 3$) transmission at a rate of 1.5 MHz

                      Probability of each ternary level = 0.333

Therefore, using Eq. (9.6) we get:

$$\eta = \frac{2 \ (\log_{10}2)^{-1} \ [0.4 \ \log_{10}(0.4)^{-1} + 0.6 \ \log_{10}(0.6)^{-1}]}{1.5(\log_{10}2)^{-1} \ 3[0.333 \ \log_{10}(0.333)^{-1}]} \times 100\% = 82\%$$

It is usual in practical systems to accept an efficiency of order 40 percent, since this readily permits the inclusion of spectrum-shaping features, etc., within the coding structure. Higher efficiency and acceptable transmission properties can normally only be obtained at the expense of more complex coding equipment. This will be obvious when we consider block codes later in this chapter.

## 9.3 ENERGY SPECTRUM SHAPING

It is appropriate to consider the objectives of energy spectrum shaping. Our analysis, as always, must begin by investigating the transmission media and its associated equipment. Typically we may observe the following characteristics:

1. Large attenuation at very low frequencies
2. Equalization is difficult, due to the physical size of the components at low frequencies
3. Large attenuation at very high frequencies (attenuation proportional to square root of transmission frequency)
4. Crosstalk between neighboring pairs increases dramatically at high frequencies

Consequently we may reasonably suppose that an ideal transmission waveform will not have a significant energy content at either high or very low frequencies.

In practice we may realize this objective by the judicious choice of three variables that are somewhat interrelated, namely:

1. The pulse shape
2. The pulse repetition rate
3. The number of logical levels that may be transmitted

Let us suppose that we wish to increase the low-frequency energy content of the coded signal. In essence our aim must be to include extra pulse transitions than would occur with a normal full-baud binary signal. We may for example adopt the *top hat*, or *di-pulse* conversion illustrated in Fig. 9-1. The associated energy spectra for these pulse types are shown in Fig. 9-2.

It is not usually possible to reduce the high-frequency energy content of the coded signal by performing the reverse process to that outlined above. Such a scheme would imply that the message signal pulses should be artificially broadened prior to transmission, and would result in an increase in the intersymbol interference (see Sec. 11.1). Thus we must think in terms of changing our other variable, the pulse repetition rate, if the high-frequency energy is to be reduced. This may be achieved by using a multilevel coding scheme that allows us to transmit the same quantity of information as the binary case, but at a reduced pulse repetition rate.

Efficient multilevel coding schemes require complex circuitry for their implementation. Consequently their use is strictly limited to high-capacity systems where the additional expense is warranted (see Sec. 9.4.4).

The calculation of the power spectral density for any given code is no easy matter, and involves several pages of complex mathematics. The classic paper on this subject

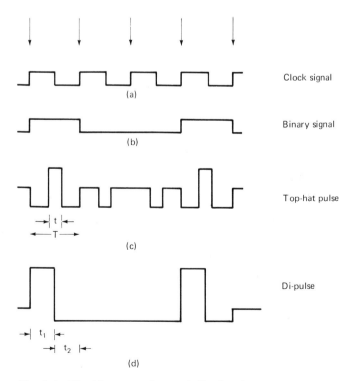

**Fig. 9-1**  The binary, top-hat, and di-pulse shapes.

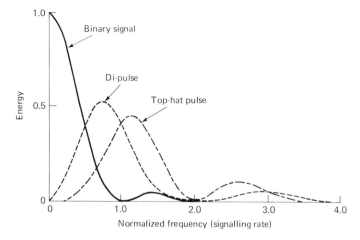

**Fig. 9-2** The relative energy spectra for different pulse types.

is by W. R. Bennett,[2] who treats the cases of random unipolar, and random bipolar signals. He obtains the result that the power spectral density $P(f)$ for a pulse amplitude modulated signal $x(t)$ is given by

$$x(t) = \sum_{n=-\infty}^{\infty} a_n g(t - nT) \tag{9.7}$$

$$P(f) = \frac{|G(f)|^2}{T} \left\{ R(0) - \overline{a^2} + 2 \sum_{K=1}^{\infty} [R(K) - \overline{a^2}] \cos 2\pi K f T \right\} \tag{9.8}$$

where $G(f) =$ fourier transform of the pulse shape
$R(K) = \overline{a_n a_{n+K}}$
$a = \overline{a_n}$ for all values of $n$

In addition to the continuous spectrum defined by Eq. (9.8), there exist discrete spectral lines $S$ at the frequencies $K/T$, given by

$$S = \frac{2\overline{a^2}}{T^2} \left| G\left(\frac{K}{T}\right) \right|^2 \tag{9.9}$$

These results are expressed in very general terms, and apply to various types of pulse sequences. The specific cases of bipolar (AMI), and unipolar coding have power spectral densities that may be deduced as discussed below.

### 9.3.1 Bipolar (AMI): Power Spectral Density Calculation

For a random signal, the probabilities of occurrence of a *mark* (logical 1), or a *space* (logical 0), are equal. Furthermore there also exists an equal probability that the mark will be positive or negative.

Thus
Probability $(a_n = 0) = \frac{1}{2}$
Probability $(a_n = +1) = \frac{1}{4}$
Probability $(a_n = -1) = \frac{1}{4}$

[2] W. R. Bennett, "Statistics of Regenerative Digital Transmission," *Bell System Technical Journal*, vol. 37, 1958, p. 1501.

There will be an equal number of positive and negative marks on average, due to the alternating action of the code translator.
Consequently,

$$\overline{a_n} = 0$$

Now $R(K)$ may be determined by evaluation for $K$ equal to 0, 1, 2, etc., separately.

$$R(0) = \overline{a_n a_n} = \overline{a^2} = \frac{1}{2}$$

$$R(1) = \overline{a_n a_{n+1}}$$

where $a_n$ and $a_{n+1}$ can assume the following combinations:

| $a_n$ | $a_{n+1}$ | $a_n a_{n+1}$ |
|-------|-----------|---------------|
| 0 | 0 | 0 |
| 0 | 1 | 0 |
| 1 | 0 | 0 |
| 1 | 1 | −1 |

Therefore,

$$R(1) = \overline{a_n a_{n+1}} = -\frac{1}{4}$$

Similarly,

| $|a_n|$ | $|a_{n+1}|$ | $|a_{n+2}|$ | $a_n a_{n+2}$ |
|---------|-------------|-------------|---------------|
| 0 | 0 | 0 | 0 |
| 0 | 0 | 1 | 0 |
| 0 | 1 | 0 | 0 |
| 0 | 1 | 1 | 0 |
| 1 | 0 | 0 | 0 |
| 1 | 0 | 1 | −1 |
| 1 | 1 | 0 | 0 |
| 1 | 1 | 1 | +1 |

Therefore,

$$R(2) = \overline{a_n a_{n+2}} = 0$$

Using the same analysis presented above $R(3)$, $R(4)$, and $R(5)$, etc., can be shown to be equal to zero.

By substitution of the above results into Eq. (9.8) we get:

$$P(f) = \frac{|G(f)|^2}{T} [\frac{1}{2} + 2(-\frac{1}{4}) \cos 2\pi fT] \qquad (9.10)$$

This may be simplified to produce the result:

$$P(f) = \frac{|G(f)|^2}{T} \sin^2 \pi fT \qquad (9.11)$$

This result defines the continuous power spectral density for AMI coding. It is pertinent to point out that discrete spectral lines cannot exist since $\bar{a}$ equals zero, and therefore $S$ defined by Eq. (9.9) must also be a zero quantity.

### 9.3.2 Unipolar Pulses: Power Spectral Density Calculation

The continuous power spectral density may be calculated for unipolar pulse sequences by repeating the analysis given above. Let us consider positive going unipolar pulses only. In this case:

$$\bar{a} = \tfrac{1}{2}$$

$$R(0) = \overline{a^2} = \tfrac{1}{2}$$

$$R(K) = \overline{a_n a_{n+k}}$$

We shall assume that all symbols are independent. Therefore,

$$R(K) = \overline{a_n a_{n+k}}$$

$$R(K) = \tfrac{1}{2} \times \tfrac{1}{2} = \tfrac{1}{4}$$

By substitution of the above results into Eq. (9.8) we get:

$$P(f) = \frac{|G(f)|^2}{T} [\tfrac{1}{2} - \tfrac{1}{4} + 2\Sigma(\tfrac{1}{4} - \tfrac{1}{4}) \cos 2\pi kfT] \qquad (9.12)$$

$$P(f) = \frac{1}{4} \frac{|G(f)|^2}{T}$$

This result defines the continuous spectrum. The existence of discrete spectral lines will be dependent on the duty cycle of the transmitted pulses. For example, if unipolar pulses of 100 percent duty cycle are employed there will be no spectral lines at any frequencies other than zero (i.e., direct current), since $G(K/T)$ equals zero. On the other hand, 50 percent duty cycle unipolar pulses will result in spectral lines at frequencies $K/T$ for $K$ odd only. The transform $G(f)$ has spectral nulls when $K$ is even.

### 9.3.3 Power Spectral Density Curves

The results obtained within Secs. 9.3.1 and 9.3.2 have been used to plot the curves illustrated in Figs. 9-3 and 9-4.

The introduction of *spectral nulls* may be used to advantage in supervising the transmission path. We have already noted (Chap. 8) that occasional transmission errors will modify the energy spectrum and will produce a large component at zero frequency. The presence of such a component may be detected and used to indicate errors. (This test assumes random data.) Furthermore the spectral nulls may be used to convey low-bit-rate supervisory data by injecting sinusoidal tones at the appropriate frequency.

The analysis presented above has concentrated on binary and bipolar signals; the only coding structures for which Eqs. (9.8) and (9.9), respectively, are valid. The high density bipolar coding schemes may be analyzed in a like manner, although the evaluation of the quantity $R(K)$ is considerably more complex. We shall see later that other code types, such as the so-called *block codes* are of a completely

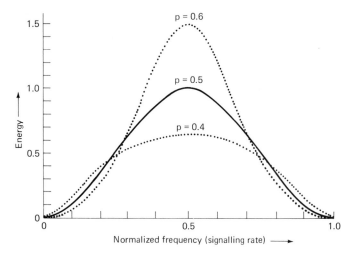

**Fig. 9-3** The power spectral density for bipolar pulses. (See Eq. (9.10), which assumes an equal probability for positive or negative marks. The probability of a mask rather than a space is also equal: i.e., $P = 0.5$. The curves also show the effect of $P = 0.4$ and $0.6$.)

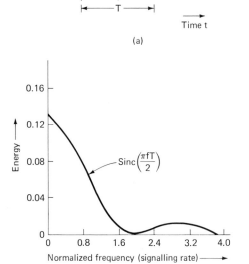

**Fig. 9-4** The power spectral density for half-width binary (unipolar) pulses.

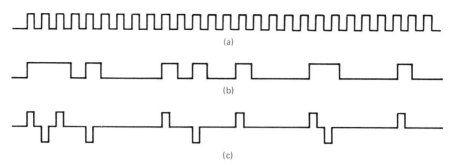

**Fig. 9-5** Binary to alternate mark inversion (AMI) code conversion: (a) clock signal; (b) binary signal; (c) AMI coded signal.

different nature. These have been analyzed in various papers, and readers interested in their properties are advised to consult the literature.[3]

## 9.4 PRACTICAL TRANSMISSION CODES

It is appropriate to briefly describe the structure of the more important transmission codes in common use today.

### 9.4.1 Bipolar, or Alternate Mark Inversion (AMI)

The coding process generates positive and negative marks, taken alternately, each occasion a binary logical 1 condition occurs within the message signal. This is shown in Fig. 9-5.

The technique was first used within the early PCM equipments developed in the U.S. by the Bell Telephone Laboratories, and is still widely used today. The code satisfies most of the requirements listed at the beginning of this chapter. For example, there is no significant dc component, the energy at low frequencies is small, and there is no error multiplication. Furthermore the bipolar signal offers a substantial improvement in immunity to crosstalk interference when compared to the unipolar signal. This advantage is of order 23 dB.[4]

The main limitation of AMI coding is the absence of timing information when the binary message signal consists of a long sequence of binary zeros. To some extent this problem can be overcome by imposing certain restrictions on the message signal. For example, in the case of the European first-order multiplexer the even numbered bits within each 8-bit sample are inverted prior to multiplexing.[5] This effectively guarantees a varying pseudorandom multiplexed binary message signal, even under idle conditions. Consequently the AMI equivalent signal will always contain adequate timing information.

Unfortunately there exist situations within a switched digital transmission network where a long sequence of binary zeros could nevertheless occur. Equipment faults

[3] A. Jessop and D. B. Waters, "4B3T, an Efficient Code for Coaxial Line Systems," *XVII Congresso Internazionale per l'Elttronica,* Roma, March 1970.

[4] M. R. Aaron, "PCM Transmission in the Exchange Plant," *Bell System Technical Journal,* vol. 41, 1962, p. 99.

[5] International Telegraph and Telephone Consultative Committee, *CCITT Orange Book,* vol. III-2, recommendations G711 and G732, International Telecommunications Union, Geneva, Switzerland, 1977.

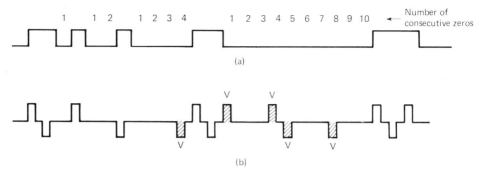

**Fig. 9-6**  Binary to high density bipolar 3 (HDB3) code conversion: (a) binary signal; (b) HDB3 coded signal.

or certain data signals[6] are possible examples of this situation.  Consequently, many telephone administrations have opted for alternative coding schemes.  The use of the AMI code in conjunction with a scrambler has also been proposed.[7]

The presence of occasional errors can be detected by observing the proper alternation of the marks.  The introduction of a single error within either a mark or a space (logical 0) will cause a discontinuity in the mark-alternation sequence.  Thus this discontinuity may be detected and regarded as indicative of a transmission error.

### 9.4.2  High Density Bipolar Codes (HDB*n*)

The coding process proposed by Croisier[8] generates positive and negative marks taken alternately each occasion a binary logical 1 condition occurs within the message signal.  However, unlike AMI coding the maximum number of consecutive zeros is limited to a value $n$.  If a series of $n + 1$ consecutive zeros does occur, the $n + 1$th zero is replaced by a mark referred to as a *violation*.  The violation mark is always inserted with the same polarity as the preceding mark.  This identifies the departure from the AMI scheme and indicates the occurrence of the violation.

In theory, if additional safeguards were not taken it would be possible for several violations to have an identical polarity, and thus introduce a dc component into the signal.  To overcome this problem, the coder continuously monitors the polarity of the violations and if two violations occur of the same polarity a double violation-insertion is made.  The first violation, in this case, occurs at the first zero in the $n + 1$ sequence, and has the inverse polarity of the preceding mark.  The second violation occurs at the $n + 1$ zero position and has the same polarity as the preceding mark (not violation).

High density bipolar coding has been widely adopted in both the United States and Europe for the transmission of first- and second-order multiplexed signals.  The most common value for $n$ is 3, in which case the code structure is referred to as HDB3.  An example of HDB3 is given in Fig. 9-6.

It is interesting to note that any HDB*n* decoder circuit can be used without restriction for the decoding of AMI signals.  Furthermore we may note that a feedback

---

[6] The CCITT recommends that data signals should be scrambled to avoid this during idle periods.

[7] See, for example, CCITT Special Study Group D, "6.312 Mbit/s Digital Interface," contribution no. 328, doc. 66, International Telecommunications Union, Geneva, Switzerland, June 1976, p. 44.

[8] A. Croisier, "Introduction to Pseudo-Ternary Transmission Codes," *IBM Journal of Research and Development*, vol. 15, 1970, pp. 354–367.

mechanism is required within both the coder and decoder circuits for the purpose of injecting violations. The delay introduced for this feedback path effectively restricts the use of HDB$n$ coding to lower-bit-rate systems. (The feedback signal may be retimed together with all other signals, but this complicates the circuitry needlessly.)

The presence of occasional errors can be detected by observing the proper alternation of the violations, since the introduction of a single error will eventually cause a discontinuity in the violation-alternation sequence. These occasional single digit errors will, however, cause more than one error in the decoded binary signal. The multiplication factor can be as high as 2.0, depending on the decoding logic used, although a more usual figure lies in the range 1.1 to 1.7 for HDB3.

### 9.4.3 Bipolar with $n$ Zero Substitution (B$n$ZS)

This coding scheme is somewhat similar to the high density bipolar translation envisaged above. The rules of alternate mark inversion are followed unless $n$ consecutive zeros occur together. Let us consider the case where $n$ equals 3.

If a block of three consecutive zeros occurs, it is replaced by the sequence $B0V$ or $00V$, where $B$ represents a pulse conforming to the AMI rule, and $V$ represents a pulse violating this rule. The $B0V$ or $00V$ pattern is selected such that the number of $B$ pulses between consecutive $V$ pulses is odd.

The bipolar with 6 zero substitution (B6ZS) code is somewhat similar. In this case a block of six consecutive zeros is substituted by the sequence $0 + - 0 - +$ if the preceding mark was positive, or $0 - + 0 + -$ if the preceding mark was negative ($+$ or $-$ represents a positive or negative mark). Codes of this type have been widely used in the United States. The B6ZS code is used at bit rates of 6.312 Mbit/s, while the B3ZS is used at the higher bit rate of 44.736 Mbit/s.[9]

### 9.4.4 The Four Binary, Three Ternary (4B3T) Code

Earlier in this chapter we introduced the concept of improved coding efficiency, at the expense of a more complex implementation. One approach suggested by A. Jessop and D. Waters of the Standard Telephone Laboratories (STL), Harlow, England, is to use a block code such as 4B3T.[10]

The coder is arranged to convert groups, or blocks, of 4 binary digits into groups of 3 ternary digits. Consequently a three-quarters reduction in the transmission rate of the coded signal is obtained.

It is clear that the binary digits represent 16 possible code words, and that these may be coded into any of 27 possible ternary combinations. Thus the design objective must be to equate each of the 16 states with a particular ternary sequence, and to do this in such a way that the transmission characteristics are optimized. The coding alphabet for the 4B3T Jessop-Waters code is shown in Fig. 9-7.

The coder selects the appropriate mode, dependent on the past history of the signal, and by reference to the state diagram shown in Fig. 9-8. Six states are specified, equivalent to the disparity counts $-3, -2, -1, +1, +2, +3$. The example shown in Fig. 9-8 will explain the operation more clearly.

Let us assume that the total accumulated disparity count at the beginning of the coding cycle is $+2$. Blocks of zero disparity have identical positive and negative modes, therefore selection is unnecessary. The binary sequence 0110 may, however,

[9] CCITT Special Study Group D, "6.312 Mbit/s Digital Interface," pp. 44, 48.
[10] A. Jessop and D. B. Waters, "4B3T, an Efficient Code for Coaxial Line Systems."

| Binary word | Ternary word | | Disparity |
| --- | --- | --- | --- |
| | Positive mode | Negative mode | |
| 0 0 0 0 | 0 − + | 0 − + | 0 |
| 0 0 0 1 | − + 0 | − + 0 | 0 |
| 0 0 1 0 | − 0 + | − 0 + | 0 |
| 1 0 0 0 | 0 + − | 0 + − | 0 |
| 1 0 0 1 | + − 0 | + − 0 | 0 |
| 1 0 1 0 | + 0 − | + 0 − | 0 |
| 0 0 1 1 | + − + | − + − | 1 |
| 1 0 1 1 | + 0 0 | − 0 0 | 1 |
| 0 1 0 1 | 0 + 0 | 0 − 0 | 1 |
| 0 1 1 0 | 0 0 + | 0 0 − | 1 |
| 0 1 1 1 | − + + | + − − | 1 |
| 1 1 1 0 | + + − | − − + | 1 |
| 1 1 0 0 | + 0 + | − 0 − | 2 |
| 1 1 0 1 | + + 0 | − − 0 | 2 |
| 0 1 0 0 | 0 + + | 0 − − | 2 |
| 1 1 1 1 | + + + | − − − | 3 |

**Fig. 9-7**   The 4B3T Jessop-Waters coding alphabet.

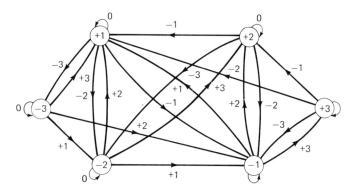

**Fig. 9-8**   The 4B3T code state-transition diagram.

be coded as 00+ (positive), or 00− (negative), dependent on the prevailing accumulated disparity count. In the example, the negative mode is selected in accordance with the state diagram for the case of an accumulated disparity equal to +2 (see Fig. 9-7). In this case the new state will be +1.

It should be noted that the state diagram contains no state corresponding to zero disparity. Therefore it is quite possible, as shown in the example, to move from state −1 to state +1 in a single step, for a ternary word disparity of +1.

The above analysis would seem to suggest that the accumulated disparity count will always lie in the range −3 to +3. In practice, however, this will not happen for two reasons:

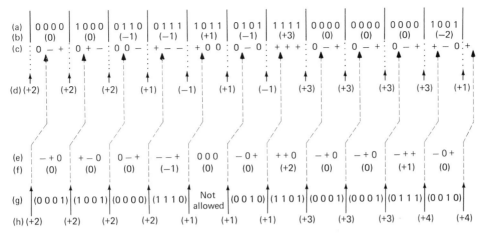

**Fig. 9-9** An example of misframed 4B3T decoding: (a) the original binary word framed in blocks of four; (b) the disparity associated with each word (see Fig. 9-7); (c) the coded equivalent ternary word (see Fig. 9-7); (d) the accumulated disparity count (initial condition: +2); (e) the misframed ternary word; (f) the disparity associated with each misframed word; (g) the retrieved binary word (see Fig. 9-7); (h) the accumulated disparity count when misframed.

1. Occasional transmission errors
2. Misframing of the block

The accumulated disparity count will very gradually move out of range when influenced by occasional errors. To overcome this problem the accumulated disparity counter must be reset each time the range is overstepped.

On the other hand, very frequent overstepping of the disparity-counter range can only be associated with misframing. An example of this is shown in Fig. 9-9h where the ternary code word shown in Fig. 9-9c is framed differently. If this were decoded a completely different binary message, as shown in brackets, would be obtained.

Therefore, in practice the overstepping rate must be continuously monitored and used to signal when reframing is necessary. The reframing operation itself is easily performed by slipping 1 ternary digit in time, until a satisfactory overstepping rate is recorded.

The practical realization of a 4B3T coder and decoder is shown in Fig. 9-10. The basic functions involved should be self-explanatory from the description given above, however, a few points are noteworthy. First, the majority of the coding circuitry operates at a frequency that is one-quarter of the binary message rate. Second, the circuit naturally performs retiming of the signal within the serialization and deserialization stages. These factors make this type of coder particularly attractive for very high-capacity systems.[11]

The 4B3T coder must normally be used in conjunction with a scrambled binary message signal. This is particularly true in the case of the European transmission hierarchy, which is based on multiplexing 4 channels at a time. A problem occurs when any multiplexer is used under-equipped, and combines 1 or 2 channels only. In such a situation the bit interleaved 4-digit sequence will contain as many as 3

[11] FACE, "800 Mbit/s Experimental Digital Transmission System," Standard SPA., Via Bodio, Milano, Italy, 1973.

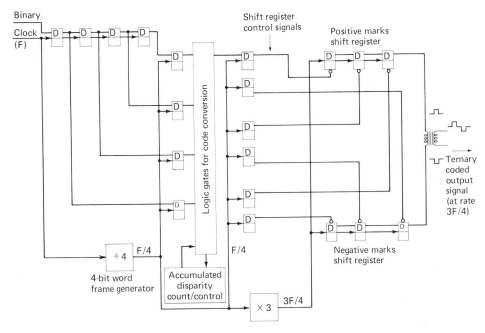

**Fig. 9-10**  Implementation of a 4B3T coder.  Edge synchronism and phasing of the clocks shown is a major problem in the design of these coders.  This is not taken into account in the diagram.

digits that remain unchanged.  The reframing mechanism cannot properly synchronize to such a signal.

### 9.4.5  The MS43 Code

The MS43 code proposed by Franaszek is similar in concept to the 4B3T Jessop-Waters code described above.  In this case 4 binary digits are coded to give a 3-digit ternary sequence; however, more than two coding modes may be selected. This feature ensures that the maximum permitted disparity count never exceeds +2 or −2.

The original[12] coding structure proposed by Franaszek defined three possible coding modes.  Here we shall consider the modified[13] form of this code which has become widely adopted within many digital transmission systems.  The coding alphabet is shown in Fig. 9-11, while the associated state-transition diagram is separately identified within Fig. 9-12.

The overall performance characteristics of the MS43 coding scheme are somewhat better than the 4B3T code; although this is achieved at the expense of increased circuitry complexity within the terminal equipments.  In particular the total pattern-induced jitter performance, and the resistance to errors caused by occasional misframing, offers some improvement for MS43 coding.  Furthermore the wandering of the

[12] P. A. Franaszek, "Sequence State Coding for Digital Transmission," *Bell System Technical Journal,* vol. 47, 1968, p. 143.

[13] L. Albertazzi, F. Beguetti, and L. Bellato, *34 Mbit/s Transmission over Microcoaxial Cable: System Philosophy and Experimental Results,* Telettra SPA., Vimercate, Italy, 1974.

| Binary words | Ternary words | | | | Disparity |
|---|---|---|---|---|---|
| | Mode 1 | Mode 2 | Mode 3 | Mode 4 | |
| 0 0 1 1 | 0 − + | 0 − + | 0 − + | 0 − + | 0 |
| 0 1 0 1 | − 0 + | − 0 + | − 0 + | − 0 + | 0 |
| 0 1 1 0 | − + 0 | − + 0 | − + 0 | − + 0 | 0 |
| 1 1 1 0 | + − 0 | + − 0 | + − 0 | + − 0 | 0 |
| 1 1 0 1 | + 0 − | + 0 − | + 0 − | + 0 − | 0 |
| 1 0 1 1 | 0 + − | 0 + − | 0 + − | 0 + − | 0 |
| 1 0 0 0 | + − + | + − + | + − + | − − − | +1, +1, +1, −3 |
| 1 0 0 1 | 0 0 + | 0 0 + | 0 0 + | − − 0 | +1, +1, +1, −2 |
| 1 0 1 0 | 0 + 0 | 0 + 0 | 0 + 0 | − 0 − | +1, +1, +1, −2 |
| 1 1 0 0 | + 0 0 | + 0 0 | + 0 0 | 0 − − | +1, +1, +1, −2 |
| 0 1 1 1 | − + + | − + + | − − + | − − + | +1, +1, −1, −1 |
| 1 1 1 1 | + + − | + + − | + − − | + − − | +1, +1, −1, −1 |
| 0 0 0 1 | + + 0 | 0 0 − | 0 0 − | 0 0 − | +2, −1, −1, −1 |
| 0 0 1 0 | + 0 + | 0 − 0 | 0 − 0 | 0 − 0 | +2, −1, −1, −1 |
| 0 1 0 0 | 0 + + | − 0 0 | − 0 0 | − 0 0 | +2, −1, −1, −1 |
| 0 0 0 0 | + + + | − + − | − + − | − + − | +3, −1, −1, −1 |

**Fig. 9-11** The modified MS43 code translation table.

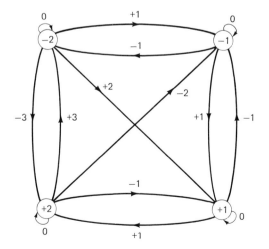

**Fig. 9-12** The modified MS43 code state-transition diagram.

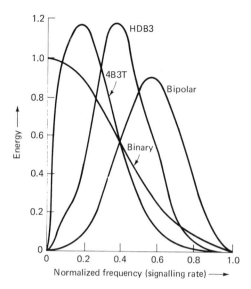

**Fig. 9-13** The energy density spectra for commonly used transmission codes. (From P. Bylanski and D. G. W. Ingram, "Digital Transmission Systems," *IEE Telecommunications Ser. 4*, Peregrinus, Stevenage, England, 1976.)

dc level is restricted to a range proportional to the disparity limits $\pm 2$, rather than $\pm 3$. This simplifies the decision threshold implementation at each repeater.

### 9.4.6   The Six Binary, Four Ternary (6B4T) Code

The 6B4T code, proposed by R. Catchpole,[14] is another example of a block code. In this case 6 binary digits must be coded to give a 4-digit ternary sequence, and therefore permits a two-thirds reduction of the transmission bit rate. This code has been proposed for use within very high-capacity digital transmission systems (e.g., 140 Mbit/s and above), where the reduction in transmission rate is an important factor.

### 9.4.7   Pair Selected Ternary (PST)

The PST code is yet another well-known example of a block code. The code translation in this case converts 2 binary digits to produce a sequence of 2 ternary digits. Therefore the transmission rate required for the coded signal remains equivalent to that of the message signal. However, the other benefits of the block-coded signal, that is low dc wander, high timing content, and the ability to detect occasional errors, continue to apply. In addition the circuitry required for the implementation of this coding scheme is considerably less than that required for the 4B3T and MS43 codes.

### 9.4.8   Other Codes

The codes listed above represent a small fraction of the large number of coding types that have been proposed over the years. The associated power spectral density curve for the most important codes in use today is illustrated in Fig. 9-13.

[14] R. J. Catchpole, "Efficient Ternary Transmission Codes," *Electronic Letter,* vol. 2, no. 20, 1975.

In conclusion it is appropriate to also identify the following coding structures that have also been adopted within many digital equipments:

1. DUOBINARY

   A. Lender, "Duobinary Coding," *Lenkurt Demodulator,* vol. 12, no. 2, 1963; idem, "The Duobinary Technique for High-Speed Data Transmission," *IEEE Transactions on Communications and Electronics,* vol. 82, 1963, pp. 214–218.

2. PARTIAL RESPONSE CODING

   F. K. Becker, E. R. Kretzmer, and J. R. Sheehan, "A New Signal Format for Efficient Data Transmission," *Bell System Technical Journal,* vol. 45, 1966, p. 755.

3. MULTILEVEL CODING

   Yukio Nakagome, Kittaro Amano, and Chuichi Ohta, *A Multilevel Code with Improved Lower Frequency Cutoff Response,* Technical Group on Communication Systems Monograph, IECE of Japan, 1967.

# 10

# Timing Extraction
# and Jitter

## 10.1 INTRODUCTION

When a regenerated clock signal varies in phase it is said to *jitter*. If we view such a signal on an oscilloscope the stable clock pattern appears to be modulated by some disturbance which is typically of low frequency (0 Hz to 1 or 2 kHz). The source of such disturbances can be attributed to one of two causes, as described below.

### 10.1.1 Pattern-Induced Jitter

Let us assume that the tuned circuit used at each repeater to generate the clock is mistuned. We may understand the effect of this by considering the tuned circuit to have a resonant frequency $f_T$, while the received data signal has an associated timing element $f_0$. In this case, if there are long gaps between the timing impulses given to the tuned circuit (see Fig. 8-6), the regenerated clock signal will drift from a frequency $f_0$ towards $f_T$, and be forced back to $f_0$ when the next timing impulse arrives. Obviously the degree of drift must be dependent on two factors:

1. The spectral density of the data signal, and consequent timing impulses
2. The degree of mistuning

The first factor gives rise to the name of this type of jitter, and has provoked considerable analysis into the investigation of transmission codes from the jitter reduction point of view. The second factor has also led to considerable analytical research, which we shall review in Secs. 10.2 and 10.3.

### 10.1.2 Waiting-Time Jitter

There is a delay between when adjustment to the reconstructed tributary clock at an asynchronous multiplexer should and does occur. This causes low-frequency clock phase shifts, and is discussed at length in Sec. 7.3.

It is clear that in a complex transmission network employing many repeaters and asynchronous multiplexers for a given signal path there will be an accumulation of both types of jitter. Consequently it is necessary to postulate the effect of jitter from the overall signal quality point of view. There will be two degradations in the signal quality:

1. IRREGULAR SPACING OF CODED SAMPLES
   In most cases the digital message is derived from an encoded analog signal.

The effect of jitter will be to cause the regenerated samples to be irregularly spaced, and consequently introduce within the baseband filtered signal a low-frequency distortion component. This is not a problem in primary PCM voice systems, but is a major problem where data transmission and wideband coded signals, such as FDM assemblies, or TV signals are involved.

### 2. IMPAIRMENT OF THE ERROR-RATE PERFORMANCE

We can identify several equipments within a transmission network that will introduce errors during periods of excessive jitter. The most sensitive mechanisms, however, must be within the regeneration process and in the capacity of the elastic store within an asynchronous multiplexer.

The regeneration process is dependent on the accuracy with which the eye is aligned with the timing signal (see Chap. 8). The effect of jitter must be to degrade the accuracy of this alignment, and hence introduce errors (see Sec. 10.5).

The elastic stores used within the tributary circuitry of an asynchronous multiplexer are typically 8 bits wide. This permits up to approximately 3 bits of jitter to be accommodated, and a jitter frequency of order 1 kHz, before the store is over- or under-flowed. If the jitter exceeds the capacity of the store, obviously errors will be introduced. It is pertinent to point out that errors are similarly introduced at the input to digital switching exchanges when excessive jitter occurs.

The practical effects of jitter on different types of digitized message signals have also been evaluated in recent years. The effect of jitter on studio quality sound signals has been evaluated by the BBC.[1] The results indicate that the higher the jitter frequency is, the more disturbing it becomes.

Black-and-white television pictures are somewhat tolerant to small levels of jitter. Color television pictures based on PAL encoding are, on the other hand, extremely sensitive to even small amounts of jitter. The resultant picture suffers badly from color hue and saturation errors.

We shall now investigate certain aspects of the jitter problems referred to above in more detail.

### 10.2  HIGH- AND LOW-$Q$ CLOCK EXTRACTION

Mistuning of the clock extraction circuit within a regenerative repeater is the most important contributor to pattern-induced jitter. The effect of mistuning is to produce a jitter related to the signal pattern density, and with a rms value of the jitter amplitude $J_A$ given by

$$J_A \, \alpha \, \sqrt{Q} \, \frac{\Delta f}{f_0} \tag{10.1}$$

where $Q =$ quality factor of the tuned circuit.

$$\Delta f = |f_T - f_0| = \text{degree of mistuning}$$

where $f =$ timing element frequency of original jitter-free signal.

---

[1] W. I. Manson, "Digital Sounds Signals: Subjective Effect of Timing 'jitter'," BBC research department, report R.D., London, England.

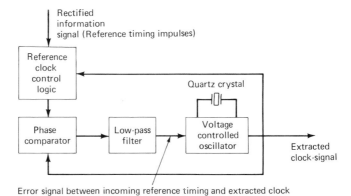

**Fig. 10-1**  An approach to high-Q timing using a PLL.

The current practice is to use both low-$Q$ (approximately 80), and high-$Q$ (approximately 1000) clock extraction circuits.  Most countries, however, have based their systems on the low-$Q$ scheme with the notable exceptions of Germany and Italy which employ high-$Q$ quartz-crystal resonators.  (Some countries in Latin America have also adopted high-$Q$ solutions.)  It is pertinent to examine the virtues of each approach.

The clock accuracy for high-$Q$ resonators is very high, and matches that of the signal element timing $f_0$.  Thus a very small degree of mistuning $\Delta f$ occurs in high-$Q$ systems.  This is accompanied by a very short decay constant, which unless special steps are taken results in the loss of the extracted clock between widely spaced reference-clock impulses.

High-$Q$ timing-recovery circuits have been developed using both *passive quartz-crystal filtration,* and *quartz-crystal controlled phase-locked loops.*  The PLL approach is the most common, and its analysis is indicative of the problems involved using high-$Q$ resonators (see Fig. 10-1).

The rectified information signal is used to synchronize the VCO.  Where gaps occur between the reference-timing impulses they are ignored by using the extracted clock to stimulate the phase comparator.  Consequently there exists a "flywheel" mechanism that keeps the extracted clock nominally referenced to the input information signal.  The time constant of this flywheel is due to the low-pass filter, which also dictates the lock time $\tau$ when inserting a new repeater, for example.  Consequently there exists a compromise between ensuring an adequate flywheel and a small lock time.

The lock time $\tau$ for a single repeater must be small (i.e., less than 1 s) in order to ensure that transmission paths comprising several repeaters $n$ can be set up rapidly. System constraints typically insist on a return-to-service condition after rectification of line faults, etc., in the order of a few seconds.  This dictates the total lock time $n\tau$ permitted for the system.  Some equipments (notably Siemens' in Germany) employ a switched dual time-constant low-pass filter to ensure both an adequate flywheel and rapid synchronization time.  This technique further complicates the circuitry but would appear to offer the perfect compromise.

Low-$Q$ timing recovery is typically achieved using a simple *LCR* tank circuit arrangement.  The approach is shown in Fig. 8-6, and no further explanation is necessary.

## 10.3  THE JITTER DUE TO A SINGLE REGENERATOR

Let us assume a clock extraction mechanism based on a low-$Q$ tank circuit. The tank circuit will be stimulated by timing impulses derived from the information signal, according to Fig. 8-6. A suitable level of timing content is ensured within the information signal by using an appropriate transmission code. Thus, a strong spectral line exists at the signal element timing frequency $f_0$. The resonant frequency of the tank circuit $f_T$ is tuned to the same value. The transfer of the tank circuit is given by

$$H(\omega) = \frac{1}{R + i(\omega L - 1/\omega C)} \tag{10.2}$$

where $R$, $L$, and $C$ = resistance, inductance, and capacitance, respectively, of the tuned circuit (serial mode).

Consequently the relative transmission of the signal element timing signal across the tank is given by

$$\frac{H(\omega)_T}{H(\omega)_0} = \frac{1}{1 + i\dfrac{\omega_0 L}{R}(\omega_T/\omega_0 - \omega_0/\omega_T)} \tag{10.3}$$

$$\frac{H(\omega)_T}{H(\omega)_0} = \frac{1}{1 + iQ\left(\lambda_C - \dfrac{1}{\lambda_C}\right)} \tag{10.4}$$

where

$$Q = \frac{\omega_0 L}{R} \tag{10.5}$$

$$\lambda_T = \frac{f_T}{f_0} = \frac{\omega_T}{\omega_0} \tag{10.6}$$

Thus, it is apparent that an effect of mistuning the tank circuit will be to introduce a static phase shift $Q$ defined by

$$\tan \theta = Q\left(\lambda_T - \frac{1}{\lambda_T}\right) \tag{10.7}$$

This effect is normally not serious in practical equipments since phase adjustment is always provided between the complete timing extraction circuitry and the retiming elements. Thus the introduction of a phase shift due to a mistuning of the tank can be compensated for.

A second more serious problem associated with mistuning is the jitter introduced by the tendency of the tank to drift towards $f_T$ when the information signal has low spectral content. Ideally an infinite $Q$ tank would settle about the strong spectral line $f_0$ of the information signal. In practice the tank does not have an infinite $Q$ and it accepts contributions for timing reference from:

1. The spectral line at $f_0$
2. The continuous spectrum immediately about $f_0$

3. Random components due to noise, and intermodulation crosstalk within the transmission path.

Within a transmission path comprising several repeaters the effects of noise and crosstalk on the jitter will be uncorrelated at each regenerator. Thus in practice their effect will be minimal. The effect of the continuous spectrum about $f_0$ will, on the other hand, introduce similar contributions to the overall jitter performance provided that the degree and direction of the mistuning remains the same at each repeater.

Analysis of the jitter magnitude against the degree of mistuning for a given information signal coding scheme is somewhat complex and beyond the scope of this book. The reader is referred to the standard texts on the subject by Manley,[2] Byrne,[3] and Bellato.[4]

## 10.4 THE ACCUMULATION OF JITTER IN A DIGITAL TRANSMISSION LINK

The accumulation of pattern-induced jitter due to several cascaded clock extraction mechanisms has been analyzed in some detail by Byrne and his colleagues at Bell Labs.[3] Later work by C. Cock of ITT,[5] based on the Byrne paper, provides the basis of discussion throughout Secs. 10.4, 10.4.1, and 10.4.2.

It is necessary to make several assumptions about the transmission mechanism. These are:

1. Each of the regenerators in the considered link is similar.
2. The pattern-induced jitter $J$ due to each section of the link is identical for all sections.
3. The pattern-induced jitter $J$ may be added at repeater $N$ to the composite signal, including jitter, from repeater $N - 1$.
4. Clock extraction is performed using a low-$Q$ tank circuit tuned to the signal-element timing rate $f_0$.
5. The rate of change of jitter is considerably less than the extracted clock rate. Consequently the jitter transfer function of the tank circuit can be represented by its low-pass characteristic.

It is also necessary to make assumptions about the behavior of the pattern itself. Two situations are important for practical purposes. The case of purely random data, which may be represented by pseudorandom test signals of order $2^{15} - 1$, and above. In addition the case of repetitive pulse-pattern blocks also needs analysis, since such signals are frequently used as test signals.

### 10.4.1 Repetitive Patterns

We consider here two test patterns, $A$ and $B$, of, say, 32 digits. Suppose pattern $A$ has a limited timing content consisting mainly of zeros, or gaps, between the timing

[2] J. M. Manley, "Generation and Accumulation of Timing Noise in PCM Systems: An Experimental and Theoretical Study," *Bell System Technical Journal,* vol. 48, 1969, p. 541.

[3] C. J. Byrne, B. J. Karafin, and D. B. Robinson, "Systematic Jitter in a Chain of Digital Regenerators," *Bell System Technical Journal,* vol. 42, 1963, p. 2679.

[4] L. Bellato and G. L. Cariolaro, "Time Jitter in Line Regenerators with Pattern Dependent Pulse Waveforms," *Alta Frequenza,* vol. 41, 1972, p. 428E.

[5] C. C. Cock, "The Accumulation of Jitter in Digital Line Links," *Institution of Electrical Engineers Colloquium on Jitter in Digital Communication Systems,* London, 19 October 1977, digest no. 1977/45.

impulses; while pattern *B* has a high timing content. In this case there will be a tendency for the extracted clock frequency to drift toward $f_T$, the resonant frequency of the tank, when pattern *A* is selected. On the other hand, if pattern *B* is selected the extracted clock frequency will approach $f_0$, the signal-element timing rate.

Let us assume that the test pattern switches from *A* to *B* and vice versa. There will be a time delay associated with each regenerator before the timing circuit has responded to the new pattern. This delay will be cumulative along the link, and may be considered as a limiting case of jitter.

Figure 10-2 shows the resultant accumulated jitter for *N* repeaters along a transmission path. It is based on the analysis of a step function, the pattern switchover, across a low-pass filter, the tank circuit. Several conclusions may be reached:

1. The jitter amplitude is directly proportional to the number of repeaters.
2. The rate of change of phase due to the step function is limited and is proportional to the effective bandwidth of the timing recovery circuit. That is inversely proportional to the *Q* factor.
3. The rate of change of phase is independent of the number of repeaters *N*. However, its duration is proportional to *N*, and is equivalent to a temporary frequency shift of the signal-element timing rate $f_0$.

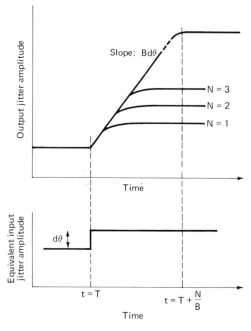

**Fig. 10-2** Jitter passed along a chain of *N* repeaters, due to a sudden pattern change at time $= t = T$. (From C. C. Cock, "The Accumulation of Jitter in Digital Line Links," *IEE Colloquium on Jitter in Digital Communication Systems,* Oct. 19, 1977, digest no. 1977/45. Copyright STL (Research), Harlow, Essex, England.)

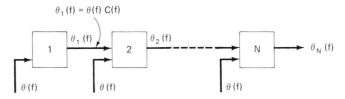

**Fig. 10-3**  Jitter accumulation for similar repeaters connected in tandem.

Normal message signals are more likely to be random than constrained to the fixed patterns envisaged here.  Nevertheless the described structure is useful for system testing, and can also be used for accurately setting up the resonant frequency of the tank.

### 10.4.2  Random Patterns

Let us assume that the message data traveling along a transmission path consisting of $N$ repeaters (see Fig. 10-3) is random.  Also let the jitter amplitude spectrum due to the pattern be represented by $\theta(f)$, and the jitter transfer function at each repeater $C(f)$.  In this case the accumulative jitter amplitude $\theta_N(f)$ at the $N$th repeater becomes:

$$\theta_N(f) = \theta(f)[C^N(f) + C^{N-1}(f) + \cdots + C(f)] \tag{10.8}$$

$$\theta_N(f) = \theta(f)C(f)\left[\frac{1 - C^N(f)}{1 - C(f)}\right] \tag{10.9}$$

The jitter transfer function $C(f)$ in the considered example is due to the low-pass response of the low-$Q$ timing recovery circuit.  Therefore,

$$C(f) = \frac{1}{1 + i\left(\dfrac{f}{B}\right)} \tag{10.10}$$

where
$$B = \frac{f_0}{2Q} \tag{10.11}$$

$$f = \text{frequency of jitter}$$

By substitution into Eq. (10.9) we get:

$$\frac{\theta_N(f)}{\theta(f)} = \frac{B}{if}\left[1 - \left(\frac{1}{1 + if/B}\right)^N\right] \tag{10.12}$$

This expression represents the normalized jitter amplitude spectrum at the $N$th regenerator.  For random data the jitter power density, that is the mean squared jitter amplitude per unit bandwidth, represents a more meaningful measure of the jitter size.  This jitter power density is given by the square of the modulus of Eq. (10.12).  Therefore,

$$\frac{J_N}{J} = \left(\frac{B}{f}\right)^2\left(1 + \frac{1}{r^{2N}} - \frac{2\cos(NP)}{r^N}\right) \tag{10.13}$$

where

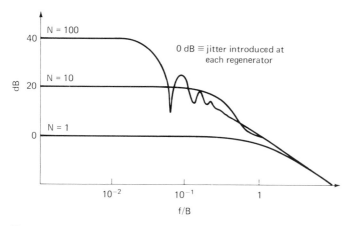

**Fig. 10-4** The normalized power density spectra for a route containing 1, 10, and 100 repeaters. (From C. C. Cock, "The Accumulation of Jitter in Digital Line Links," *IEE Colloquium on Jitter in Digital Communication Systems,* Oct. 19, 1977, digest no. 1977/45. Copyright STL (Research), Harlow, Essex, England.)

$$r^2 = 1 + \left(\frac{f}{B}\right)^2 \tag{10.14}$$

$P=$ power density contributed by each repeater

$P= \arctan (f/B)$

The normalized power density spectra for a route containing 1, 10, and 100 repeaters have been plotted using Eq. (10.14) in Fig. 10-4. At very low frequencies the jitter power density is proportional to $N$ and therefore very low frequency jitter components tend to accumulate with an amplitude directly proportional to $N$. The troughs seen in the spectra for a large number of repeaters are of no practical significance. These are the result of amplitude additions across a large number of components, with differing phase, at each point along the abscissa.

The rms amplitude $\theta_N$ after $N$ repeaters can be shown to be given by

$$\theta_N = \sqrt{\tfrac{1}{2} JB \left( N - \tfrac{1}{2} \left\{ \frac{(2N-1)!}{4^{N-1}[(N-1)!]^2} \right\} \right)} \tag{10.15}$$

It is convenient to normalize this expression with respect to the rms jitter contributed by each repeater. This has been done, and the result plotted against the number of repeaters $N$, within Fig. 10-5. Analysis of the curve leads us to the important conclusion that for a large number of repeaters:

Accumulated rms jitter amplitude $= \alpha \sqrt{N}$    (correlated jitter source)    (10.16)

This result may be explained by simple amplitude addition of the low-frequency components, while filtration and vectorial addition has the major influence on the high-frequency components. The result dictated by Eq. (10.16) holds for jitter due to correlated sources, that is due to random pulse-code patterns. It is pertinent to analyze the case of uncorrelated sources of jitter.

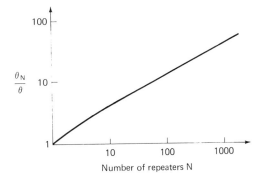

**Fig. 10-5** Normalized rms jitter versus the number of tandem repeaters. (From C. C. Cock, "The Accumulation of Jitter in Digital Line Links," *IEE Colloquium on Jitter in Digital Communication Systems,* Oct. 19, 1977, digest no. 1977/45. Copyright STL (Research), Harlow, Essex, England.)

Jitter due to uncorrelated sources may be expected to add on a purely power basis. Therefore we may expect, following a similar analysis[6] to that given above:

$$\text{Accumulated rms jitter amplitude} = \alpha \sqrt[4]{N} \quad \text{(uncorrelated jitter sources)} \quad (10.17)$$

## 10.5 JITTER TOLERANCE WITHIN REGENERATORS (ALIGNMENT JITTER)

We have already noted that the error-rate performance of the regeneration process is likely to be significantly impaired during conditions of severe jitter. It is pertinent then to ask, What are the limiting conditions, or maximum level of jitter, that can be tolerated by the regeneration process?

We have noted in Chap. 8 that ideally the information eye will always be interrogated at its center. This ensures error-free transmission. In practice however, the extracted clock signal used to interrogate the eye will jitter about the center, and the information eye will itself jitter about some mean position. The limiting case for error-free interrogation requires the peak-to-peak amplitude of the relative jitter between the eye and extracted clock to be less than the effective eye width $W$, as shown in Fig. 10-6; that is:

$$D - d \le W \quad (10.18)$$

where $D$ = jitter amplitude of information eye (peak to peak)
$\quad d$ = jitter amplitude of extracted clock (peak to peak)
$\quad W$ = effective eye width

Then, for sinusoidal jitter, and simple low-$Q$ clock extraction, we get from Eq. (10.10):

$$d = C(f)D \quad (10.19)$$

[6] O. E. de Lange, "The Timing of High-Speed Regenerative Repeaters," *Bell System Technical Journal,* vol. 37, 1958, p. 1455; and H. E. Rowe, "Timing in a Long Chain of Binary Regenerative Repeaters," *Bell System Technical Journal,* vol. 37, 1958, p. 1543.

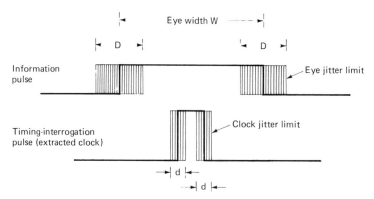

**Fig. 10-6**  Binary eye and the associated clock, with jitter.

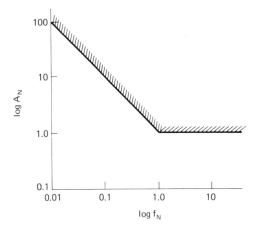

**Fig. 10-7**  Idealized jitter amplitude tolerance versus jitter frequency. (From C. C. Cock, A. K. Edwards, A. Jessop, "Timing Jitter in Digital Line Systems," *IEE Conference Publication no. 131, Telecommunication Transmission,* Sept. 1975. Copyright STL (Research), Harlow, Essex, England.)

$$d = \frac{D}{1 + i(f/B)} \qquad (10.20)$$

By substituting Eq. (10.20) into (10.18) and rearranging terms we get:

$$D \le \left| W\left(1 - \frac{i}{f/B}\right) \right| \qquad (10.21)$$

Therefore,

$$D_N \le \sqrt{1 + 1/f_N^2} \qquad (10.22)$$

where $D_N = D/W$ (normalized jitter amplitude, peak to peak)
$f_N = f/B$ (normalized jitter frequency, peak to peak)

The above has been plotted in Fig. 10-7, where it is apparent that the jitter tolerance exhibited by the low-$Q$ clock extraction mechanism envisaged above follows one of

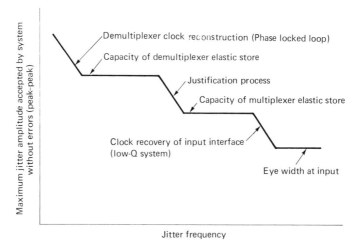

**Fig. 10-8** Jitter tolerance limits due to various processes within a digital transmission system. (From C. C. Cock, "Jitter Tolerances in Digital Equipment," *IEE Colloquium on Jitter in Digital Communication Systems,* Oct. 19, 1977. Copyright STL (Research), Harlow, Essex, England.)

two characteristics, dependent on frequency. At low jitter frequencies, below $f_N$, the timing recovery circuits tend to track the phase excursions. However, at frequencies above $f_N$ there is no dependence on the jitter frequency, and the tolerance to jitter is determined exclusively by the eye width $W$. The actual transition point between the two characteristics is dependent on several factors, namely:

1. The low-pass characteristic, or bandwidth, of the clock extraction circuitry (regenerative repeaters and line interface circuits)
2. The phase-locked loop associated with dejustification units (see Chap. 7)
3. The justification process of dependent asynchronous multiplexers (see Chap. 7)

This analysis has considered exclusively sinusoidal jitter, that is the extracted clock behaves as if phase-modulated by a sinusoid. The extension of the arguments above to the practical case, assuming pattern-induced jitter due to a transmission path consisting of several repeaters, has been performed by C. Cock.[7]

### 10.6 JITTER TOLERANCE WITHIN COMPLETE TRANSMISSION SYSTEMS

We have considered above in some detail the sensitivity of clock extraction mechanisms to jitter. We must also consider the jitter-sensitive processes that occur within asynchronous multiplexers. Those are:

1. The multiplex justification process
2. The demultiplex clock recovery mechanisms (phase-locked loop)

---

[7] Cock, "Jitter Tolerances in Digital Equipments," *Institution of Electrical Engineers Colloquium on Jitter in Digital Communication Systems,* London, 19 October 1977, digest no. 1977/45.

The jitter sensitivity of a phased-locked loop has characteristics similar to those of the simple clock extraction mechanism envisaged in Sec. 10.5.

The multiplex justification process may be considered as a mechanism that samples the clock frequency of the tributary signal. Thus the maximum jitter frequency that can be followed cannot be much higher than the nominal sampling rate. We may expect therefore an upper limit on the peak-to-peak jitter amplitude to be imposed by the capacity of the multiplex justification store. At very low jitter frequencies, below the justification rate, the justification process itself will accommodate some of the jitter. In this case the demultiplex dejustification store and the phase-locked loop are the limiting factors.

Tests performed at Standard Telephone Laboratories suggest that the jitter tolerance characteristic for an asynchronous digital multiplexer behaves as shown in Fig. 10-8. There are six separate frequency-dependent limitations to the minimum tolerance to jitter accepted by an asynchronous digital multiplexer. These are:

1. The phase-locked loop
2. Demultiplex dejustification store
3. The justification process
4. Multiplex justification store
5. Clock recovery at line interface circuits (see Fig. 10-7)
6. Information eye width (see Fig. 10-7)

# 11

# Equalization

We have already noted in Chap. 8 that equalization within a cable line system performs two main functions. These are:

1. Compensation for certain frequency-dependent cable parameters
2. Simple shaping of the transmitted pulses

We shall now investigate the waveforms occurring at the system input $S_i(t)$ and equalizer output $S_0(t)$, as shown in Fig. 11-1. Both the channel input and equalizer output signals have finite transition times, yet their pulse shapes differ significantly. This may seem somewhat surprising, since intuitively we may have anticipated that the preamplification/equalization function completely compensated for all pulse-shape degradations introduced by the transmission medium. This is not necessary since we are only interested in the pulse condition at the sampling instants. At all other times the regeneration process ignores any degradations of the pulse shape.

The equalization process is one of compromise. On the one hand, the noise level may be minimized by choosing a pulse shape that exhibits a slow roll-off. This maximizes the effect of intersymbol interference, as shown in Fig. 11-2. Conversely a pulse shape that exhibits a sharp roll-off improves the intersymbol interference

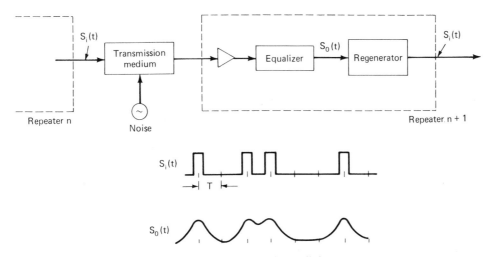

**Fig. 11-1** The basic model of a repeated transmission link.

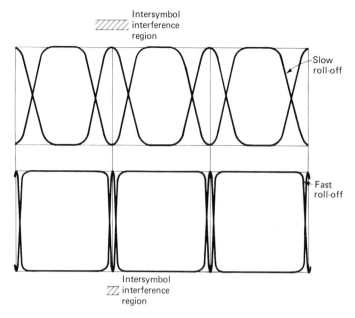

**Fig. 11-2**  The effect of pulse roll-off on intersymbol interface.

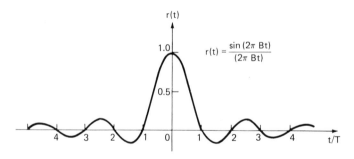

**Fig. 11-3**  The sinc function impulse response.

performance at the expense of an increased noise level; since a sharp roll-off characteristic implies increased bandwidth and hence increased noise.

The message signal may be represented by a sequence of discrete pulses $a_1$, $a_2$, $a_3$, . . . , $a_n$ that occur at finite intervals, spaced $T$ seconds apart. In this case, if we consider the degree of pulse shaping imposed by the equalizer to be fully described by the transfer function $r(t)$ we get:

$$S_0(t) = \sum_{-\infty}^{\infty} a_n r(t - {}_nT) \qquad (11.1)$$

Let us examine the function $r(t)$ in more detail. In order to ensure error-free interpretation of the message pulses we may specify that $r(t)$ be zero at all sampling instants $nT$, except for $n = 0$. (Partial response systems permit a finite value of $r(t)$ at sampling instants.) If we additionally, in the interests of noise reduction, restrict the bandwidth of the equalizer to the minimum level as defined by Nyquist, we obtain a transfer function given by:

$$r(t) = \frac{\sin(2\pi Bt)}{2\pi Bt} \qquad (11.2)$$

where $B$ = minimum bandwidth given by Nyquist theorem
$\quad B = \frac{1}{2}T$

The pulse shaping defined by Eq. (11.2) is of course that given by the sinc function referred to in Chap. 3, and redrawn in Fig. 11-3.

In choosing an appropriate pulse shaping characteristic within a practical system we need to also consider the pulse level at positions slightly displaced from the sampling instants. In this way, we can establish the likely effect of jitter or other timing imperfections.

The sinc function converges to zero very slowly. Consequently we may expect significant contributions from the tails of preceding pulses if the sampling instant is displaced slightly about its nominal position. For this reason the sinc pulse shape is rarely used in practical systems.

A more suitable pulse shape for practical systems is given by the so-called *raised-cosine* impulse response given by[1]

$$r(t) = \frac{\sin 2\pi(b+B)t \cos 2\pi bt}{\pi t[1 - (4bt)^2]} \qquad (11.3)$$

where $b$ = bandwidth of equalizer
$\quad B$ = Nyquist bandwidth

The raised-cosine response offers a smooth, and rapid convergence to zero, together with zero intersymbol interference at the sampling instants (see Fig. 11-4). This is achieved at the expense of requiring additional bandwidth compared to the minimum

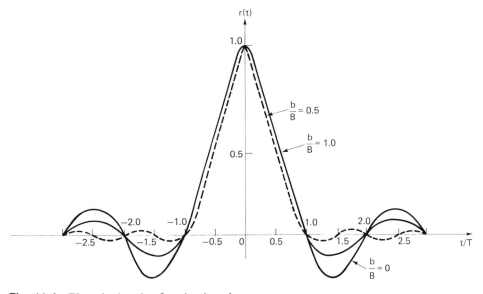

**Fig. 11-4**   The raised-cosine function impulse response.

[1] F. Haber, "Rapidly Converging Sample Weighting Functions," *IEEE Transactions on Circuits and Systems*, CS12, 1964, pp. 116–117.

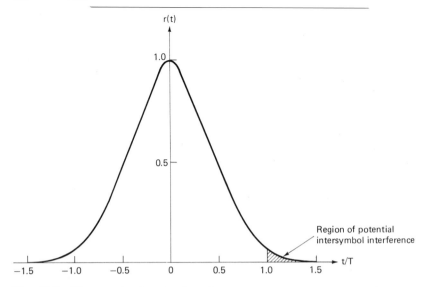

**Fig. 11-5**  The gaussian function impulse response.

defined by Nyquist.  Practical systems typically employ an excess bandwidth in the range 15 to 100 percent above the Nyquist value.

Clearly, in practical systems that are likely to be affected by timing and amplitude irregularities, a compromise must exist between the chosen convergence characteristic and the equalizer bandwidth dictated by that characteristic.  Furthermore as soon as we permit the sampling moment to be defined as a region rather than as an instant of time, it is pertinent to question the need for imposing the zero intersymbol interference criterion at the instant $nT$.

Another practical waveform that has small, yet finite, intersymbol interference at the sampling instants is given by the gaussian function defined by

$$r(t) = \exp\left(-nt^2/T^2\right) \tag{11.4}$$

The response associated with the gaussian function is illustrated in Fig. 11-5.  The convergence to zero is both rapid and smooth, while the interference at $n = 1$ is minimal.

## 11.1  PARTIAL RESPONSE SYSTEMS

The equalization schemes described above aim to optimize the pulse characteristics, such that minimum intersymbol interference occurs at the sampling instants.  Equipments based on this criterion have been successfully designed and installed in both the United States and Europe.  In the mid-sixties the concept of *partial response* was introduced by A. Lender of General Telephone and Electric (GTE), and has since been widely adopted within equipments in the United States.

Partial response systems deliberately allow intersymbol interference to occur in a controlled manner.  By choosing an appropriate coding scheme contributions will (or will not) be detected from adjacent pulses at each sampling instant.  At the receiving terminal the resultant signal is uniquely decoded by virtue of the correlative code structure chosen.

The basic concept behind correlative coding will become apparent by analysis of the duobinary, or class 1, partial response coding structure applied to binary transmission. Let us assume once again a sequence of $a_1, a_2, a_3, \ldots, a_n$ pulses, transmitted at $T$ seconds apart over an ideal rectangular low-pass channel of Nyquist bandwidth $2/T$. Let the pulse stream be applied to the digital filter shown in Fig. 11-6. It is clear that every pulse at the input $a_n$ will be interfered by the previously received pulse $a_{n-1}$.

$$A_n = a_n + a_{n-1} \tag{11.5}$$

Now, a delay element has a transfer function equivalent to $\exp(-j2\pi fT)$. Consequently, the digital filter exhibits a transfer function equivalent to $= 1 + \exp(-j2\pi fT)$. The total transfer function $H(f)$ of the channel, due to the cascade of the digital filter and the ideal Nyquist channel $N(f)$, becomes:

$$H(f) = N(f)\,[1 + \exp(-j2\pi fT)] \tag{11.6}$$

$$H(f) = N(f)\,[\exp(j\pi fT) + \exp(-j\pi fT)]\,\exp(-j\pi fT) \tag{11.7}$$

$$H(f) = 2N(f)\cos(\pi fT)\exp(-j\pi fT) \tag{11.8}$$

The corresponding impulse response $h(t)$ is shown in Fig. 11-7. It is immediately clear that only the samples occurring at times 0 and $T$ are allowed to interfere, since at all other sampling instants zero intersymbol interference occurs. This ability to

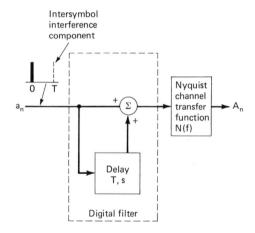

**Fig. 11-6** Partial response equalization scheme.

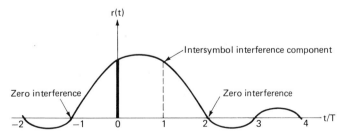

**Fig. 11-7** Duobinary function impulse response.

permit interference, but only in controlled amounts, is key to the partial response approach.

Detection of the original signal is achieved by passing $A_n$ through a mechanism that performs the inverse operation of the digital filter at the transmitter. In this way the value $a_{n-1}$ once decided may be substracted from $A_n$ to reveal the most recently received pulse $a_n$. This demodulation technique does have the drawback that any error in $a_{n-1}$ is likely to affect the proper detection of $a_n$. A technique for avoiding this error propagation was proposed by Lender, and is referred to as *precoding*.[2]

[2] A. Lender, "The Duobinary Technique for High-Speed Data Transmission," *IEEE Transactions on Communications and Electronics,* vol. 82, 1963, pp. 214–218.

# TOMORROW'S TELECOMMUNICATION SYSTEMS

# 12

# Future Trends in Telecommunication

## 12.1 INTRODUCTION

This book has dealt mainly with PCM multiplexing and digital transmission as currently applied within today's worldwide telephone networks. The described systems have been phased in with great fanfare, yet at small production equipment volumes since the early 1970s. During the last few years, however, production volumes of PCM transmission equipments have dramatically increased, and more importantly the development phase for several major PCM digital switching schemes is nearing completion.

On another front, the telephone subscriber, a hand-calculator age consumer, has come to expect more from the subset itself, and the exchange to which he or she is connected. In the telephone subset a host of facilities are already becoming available: last-number redial, station number memory, dialed-number optical readout, call charge automatically displayed, and so on. In the exchange, and private automatic branch exchanges (PABX), stored program control (SPC) switching machines have permitted a further range of facilities to be offered to the consumer: follow-me, long distance number blocking, detailed invoicing, abbreviated dialing, etc. The commercial implications of this activity are enormous for both the end-equipment manufacturer and the dependent industries such as the semiconductor suppliers. These facilities can only be realistically offered within a digital environment, and this has naturally led to a worldwide demand for PCM codecs in million units plus quantities.

The semiconductor industry has reacted to this new, potentially large market by becoming suppliers of dedicated integrated circuits, and has moved away from their traditional role of logic-gate peddlers. At the time of writing many of the functions described in this book have become available as single integrated circuits (see Table 12-1). Thus, the end-equipment manufacturer can already buy a circuit function, where the trade-offs and compromises between practice and theory that have been identified throughout this text have already been made. Equipments will become cheaper, smaller, consume less power, and require less labor to fabricate. We may as a consequence identify several trends for the future:

1. A move toward PCM multiplex equipments based on single-channel codecs, rather than the shared codec approach considered in Chap. 6.

251

**Table 12-1  Availability of Integrated Circuit Devices at the End of the Seventies**

| Transmission PCM multiplex | Transmission line | Switching systems |
|---|---|---|
| Subscriber line interface | | Subscriber-line interface |
| Single-channel filter | | Single-channel filter |
| Shared codecs | | Shared codecs |
| | | Single-channel codecs |
| HDB3 coders | HDB3 coders | HDB3 coders |
| | Justification circuits | Crosspoint |
| | Line supervision devices | |
| | PCM repeaters | |

2. Low-cost PCM multiplexers being used for transmission, switching, PABX, and subscriber circuits.
3. The digital telephone subset, based on a low-power codec ICs within the telephone set apparatus.

Technology advances have also enabled researchers to improve the transmission media itself.  Optical transmission through glass fibers is now technically feasible.  Consequently, it is already clear that as cheaper and better quality fibers become available transmission methods will radically change.

The first attempt at optical transmission began at Bell Laboratories in the 1960s where Dr. Mayo experimented with heated air lenses.  Later work in the sixties, notably by Dr. Maurer (Corning Glass, New York) and Dr. C. Kao of STL (ITT research laboratories, United Kingdom), produced optical fibers that although of high attenuation (1000 dB per km!) were nevertheless flexible, lightweight, and potentially low-cost.  This initial work proved the feasibility of optical communication.

Considerable improvements in optical fiber design occurred during the seventies.  Consequently at the start of this decade several first-generation optical communication links were in service worldwide.  The trend toward optical transmission will undoubtedly continue for reasons of cost and ease of installation.

In other areas researchers have considerably refined the techniques of analog-to-digital conversion, and the methods of transmitting the digital signal.  As a result, today it is possible to communicate many more speech channels down a voice-grade cable than the 24 or 30 considered in the text of this book.  In particular the following advances in digital technique have occurred during recent years:

1. Correlative coding and duobinary transmission have increased the data handling capabilities of the Bell T1 carrier from 24 to 48 channels.  This has been a significant development within the United States, pioneered largely by Dr. Lender of GTE.[1]
2. Adaptive equalization techniques originally developed for data transmission purposes permit the equalization characteristics of each repeater to be tailored to the particular line in use.  In this way we may push the transmission rate, or repeater separation, to limits we could not normally achieve with a nonadaptive scheme.
3. Predictive quantization and speech processing techniques have already enabled

---

[1] A. Lender, "The Duobinary Technique for High-Speed Data Transmission," *IEEE Transactions on Communications and Electronics,* vol. 82, 1963, pp. 214–218; and idem, "Correlative Level Coding for Binary Data Transmission," *IEEE Spectrum,* vol. 3, 1966, pp. 104–115.

significant reductions in the storage capacity required to contain voice messages. The Texas Instruments Speak-and-Spell toy[2] is a case in point. Obviously such techniques have significant implications for speech transmission in the future.

The trend toward adopting the above techniques will in my view be different in Europe and the developing countries compared to the United States. The approach within the U.S. telephone network has historically been to provide the lowest cost link between two points without any particular regard for hierarchy or standardization. This is evidenced by the variants of the Bell T1 systems, albeit upgrades, and the multiplicity of higher-order transmission schemes within the United States. Consequently I would believe that these new techniques will be most readily incorporated within end-equipment designs in the United States.

In Europe there is a much stronger adherence to an agreed hierarchy as laid down by the Central European Post and Telegraphy Organization (CEPT) or CCITT. Consequently it will be a much slower process, if not an impossible one, to incorporate any new techniques that impact other equipments already in service within the network. This will always be true in Europe unless the cost impact is very significant, as was the case in the PCM versus FDM battle.

The trends outlined above would seem to suggest that the digitization of the telephone network will continue for the foreseeable future. There are, however, several questions that remain unanswered; for example: What will change within today's existing end-equipments? What new equipments will be demanded by the evolving network? What will be the impact of advances in semiconductor and optical fiber technology? It is perhaps the last of these questions that holds the key to the future.

We shall now review in more detail the advances in technology that have occurred in recent years, and consider what the likely impact on telecommunications could be. In particular we shall consider the marriage of switching and transmission functions within the same equipment.

The different areas of technology development will be considered within the following sections:

## 12.2 INTEGRATED DEVICES
### 12.2.1 Integrated Codecs
### 12.2.2 Integrated Filters
### 12.2.3 Integrated Subscriber Line Interface Circuits

## 12.3 DIGITAL SWITCHING
### 12.3.1 Stored Program Control
### 12.3.2 Local Concentrators

## 12.4 DIGITAL TELEPHONE SUBSETS

## 12.5 OPTICAL COMMUNICATION
### 12.5.1 Optical Repeater
### 12.5.2 Coding within an Optical System

---

[2] Staff writer, "Single Silicon Chip Synthesizes Speech in $50 Learning Aids," *Electronics,* vol. 51, 1978, pp. 39–40; and R. Wiggins and L. Brantingham, "Three-Chip System Synthesizes Human Speech," *Electronics,* vol. 51, no. 18, 1978, pp. 109–116.

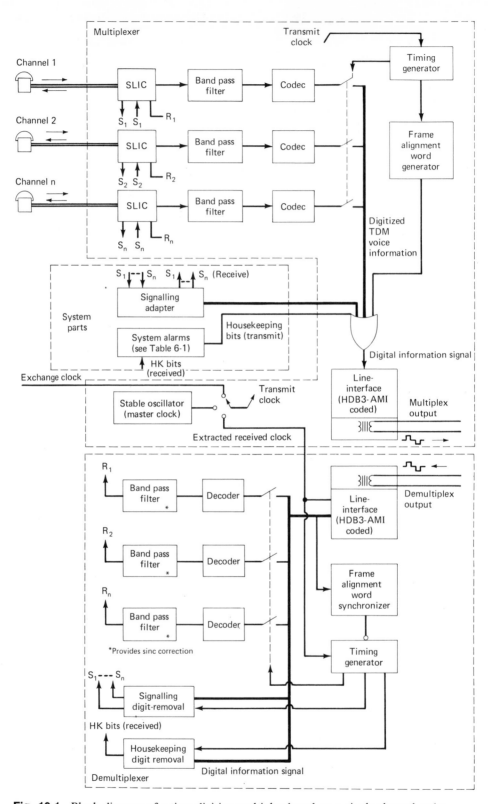

**Fig. 12-1** Block diagram of a time division multiplex based on a single-channel codec.

The impact due to advances in semiconductor technology will now be considered.

## 12.2 INTEGRATED DEVICES

### 12.2.1 Integrated Codecs

The codecs used in the early TDM equipments were formed using discrete components, and operated at high speed. One coding unit, referred to as a *shared codec,* handled several voice channels which were time division multiplexed as PAM signals (see Fig. 6-12).

Another approach based on several *single-channel codecs* is shown in Fig. 12-1. In this case a conceptually simpler codec, capable of handling 1 channel only, is used to digitize each voice channel separately. The TDM signal is then obtained by straightforward multiplexing of the digitized voice signals.

A cost and power consumption comparison between the two approaches tends to favor the shared codec solution, since the cost and power needs of the shared approach are typically less than the sum of several single-channel units, as shown in Fig. 12-2. Consequently we may ask, Why are integrated single-channel codecs so important for tomorrow's telecommunication systems, and why have these devices stimulated so much interest within the industry?

The single-channel codec is directed at switching equipments rather than transmission. The channel board identified within Fig. 12-1 significantly facilitates the switching process by providing purely digital signals which may be:

1. Deposited into a binary store (memory)
2. Interpreted by a microprocessor
3. Routed via logic gating

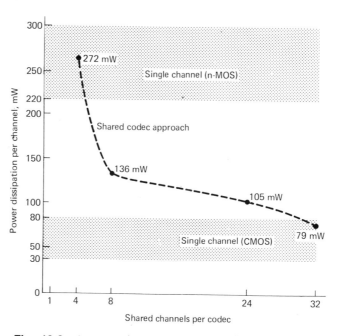

**Fig. 12-2** A comparison between single-channel and shared codec power consumption performance.

4. Transmitted over medium distances within an exchange environment without signification degradation

The main advantage then of the single-channel codec is that it permits a more rapid digitization of the analog voice signal in order to exploit the mechanisms (1) to (4) above. The eventual limit of this trend will be to include a codec within the subset instrument itself (see Sec. 12.4).

The practical integration of a single-channel codec is naturally very dependent on the process technologies available to each semiconductor manufacturer. The first generations of codecs emerged from specialist manufacturers who typically lacked a wide spectrum of process technologies. Consequently, early designs were compromises between end-equipment needs, commercial constraints, and technology availability. We may identify the most desirable features for a single-channel codec as:

1. *Small bar size, or chip area:* The bar size is directly related to the process yields, and the number of good devices per slice. Thus a small bar size ensures low cost and a producible device in large quantities. The desirable maximum bar size is related in addition to the complexity of the process (i.e., number of mask steps, etc.), and how well the manufacturer can maintain process control.

2. *Low power consumption:* The lowest possible power consumption is required for both the operational and standby conditions.

   *Note:* Typically, low power dictates a complementary metal oxide semiconductor (CMOS) process, while small bar area is favored by an *n*-type metal oxide semiconductor approach. Thus the question needs to be asked: Low power, at what price?

3. Minimum number of power supplies, with voltage and tolerances equivalent to digital logic requirements.

4. Asynchronous or synchronous operation between transmit and receive directions.

5. Minimum number of external componentry. The integration will ideally include the following features:

   *a.* A voltage reference to define the companding curve accurately.

   *b.* An *auto zero* feature that sets the midpoint of the companding curve precisely.

6. Performance equivalent to, or better than, that defined for the A and $\mu$ law multiplex channel banks.

7. *High reliability:* The process design rules, and the selection of the process plus packaging technologies must ensure that the probability of a device failure within the working life of the codec is very small. This parameter is typically measured by telecommunication engineers in FITS [failure in $10^9$ hours (h)] at some elevated temperature (e.g., 55°C). (Semiconductor manufacturers specify reliability in terms of the percentage of failures per $10^3$ at some elevated temperature.) A design goal for most single-channel codecs has become that 1 to 10 FITS maximum can be tolerated. This is one to two orders of magnitude better than that required by commercial consumer-grade electronic devices.

8. The digital writing and reading rates should be compatible with the channel groups used within the switching machines (e.g., 24-channel, 1.5 Mbit/s; 30-channel, 2.0 Mbit/s; 48-channel, 3.0 Mbit/s; 60-channel, 4.0 Mbit/s). It should be noted that only the output ports need to have the capability of working at these elevated bit rates, all other coding functions work at the single-channel rate (normally 64 kbit/s).

**Table 12-2  A Comparison between Several First-Generation Single-Channel Codecs**

| Parameter | Codec 1 | Codec 2 | Codec 3 | Codec 4 | Codec 5 | Codec 6 |
|---|---|---|---|---|---|---|
| Bar size, mm$^2$ | 22.0 | 32.0 | 34.0 | 19.5 | 36.0 | 19.2 and 16.4 |
| Technology | n-MOS | CMOS | CMOS | n-MOS | I$^2$L | CMOS |
| No. of masks | 9 | 10 | 7 | 6 | 10 | |
| Power consumption | | | | | | |
|   Operational | 220 | 30 | 80 | 300 | 375 | 45 |
|   Standby | 110 | | 0.5 | | 50 | |
| Power supply required, V | +12, +5, −5 | +5, −5 | 15 | +9, +5 | +12, −12, +5 | +7, −7, +5 |
| General comments | Small bar | Large bar | Large bar | Very high power | Very high power | 2-chip solution |
| | High power | Very low power | Medium power | | | |
| | V ref. on chip | No V ref. | Single power supply | | | |

9. Minimum system interfaces between the codec, the filter, the subscriber line interface circuitry (SLIC), and the switching machine or multiplexing circuitry.

The usual block schematic for the channel-level circuitry defines the individual units uniquely interconnected, as shown in Fig. 12-1. For switching systems in particular, however, there are many more interconnections that define such diverse functions as *power down, off hook, time-slot allocation,* and so on. We may expect during this decade that semiconductor manufacturers will tend to provide total system solutions to the channel-level circuitry, where the codec is only one of the required modules.

A comparison between several early designs of single-channel codec is given in Table 12-2. The divergence of approach is self-apparent.

### 12.2.2   Integrated Filters

The requirements for a channel filter have been discussed within Chap. 3. Typically, practical constraints such as the rejection of noise due to switched-mode power supplies, mains hum (50 Hz, 60 Hz), or even low-frequency signaling tones, need to be taken into account also. We may summarize the basic specification needs of the channel filter as:

1. Low-pass filter restricting the message bandwidth, in accordance with the Nyquist sampling theorem (see Chap. 3)
2. Band-pass characteristic (optional)
3. Equalization due to $\sin x/x$ in receive direction only (see Chap. 3)
4. Rejection of 30 to 40 dB at mains frequency, switched-mode power supply frequency, sampling frequency, and the related harmonies
5. Nominally flat response throughout the bandwidth of the message signal
6. Minimal propagation delay distortion

We may think of integrating a filter using one of the following techniques:

1. Charged-coupled device (CCD) technology
2. Switched capacitance technology
3. Digital signal processing techniques

The CCD approach is best suited to high-frequency applications, where it has been successfully employed within radio and radar equipments. At voice frequencies it is difficult to realize anything other than a low-pass response, and then only with a high degree of ripple. Thus, the goal of a nominally flat band-pass characteristic can only be met with great difficulty, using CCD techniques.

The attraction of the signal-processing approach is that the operation is digital, and can therefore be realized using simple high-speed digital devices. The drawback is, using today's technologies, a high power consumption and a large bar size.

Most practical filter designs currently available employ the switched-capacitance approach. This technique is illustrated within Fig. 12-3, where it is compared to the conceptually similar single-pole $RC$ filter. An $RC$ filter produces a transfer characteristic dependent on the decay of charge through a resistor. In a similar way switched capacitance filters permit the charge across a capacitor $C1$ to be slowly discharged; in this case by adding capacitors $C2$, $C3$, and so on, after predetermined digitally timed intervals have elapsed. Thus, a transfer characteristic is produced which has a response determined by the switching intervals and the change of capacitance value at each switching moment.

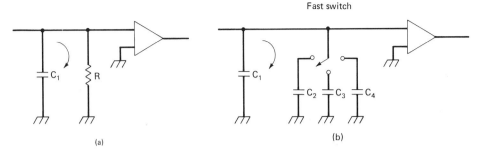

**Fig. 12-3** Comparison between (a) simple $RC$ filtration and (b) the switched capacitance equivalent.

There are several practical considerations that must be borne in mind when designing switched-capacitance filters. The capacitors $C1$, $C2$, etc., must remain stable, and maintain a fixed relationship between each other. In addition the individual capacitance values must exhibit a linear dependence on the supply voltage, or the power-supply noise rejection capability will be seriously impaired. These considerations largely determine the semiconductor technology employed.

### 12.2.3   Integrated Subscriber Line Interface Circuits

The function performed by the subscriber line interface circuit (SLIC) is commonly referred to as BORSCHT,[3] a useful acronym describing the operations involved:

B = Battery and power feeding to the subset microphone

O = Overvoltage protection

R = Ringing signal generation for the subset bell

S = Supervision of the subscriber line

H = Hybrid: The term given to the device that permits two-way communication over a single pair cable

T = Test access to the subscriber line

The classical approach to realizing the SLIC has adopted coils (hybrid), relays (supervision, test), and a carbon block (overvoltage), as shown in Fig. 12-4. The cost of implementation using these antiquated devices lies in the region of 30 to 70 U.S. dollars, but worse, the equipment space consumed is considerable. Consequently numerous *attempts* have been made to integrate one or more of the BORSHT functions.

The hybrid function can be most easily integrated using readily available semiconductor linear technology. The ringing signal, on the other hand, forces a requirement for a technology capable of withstanding 150- to 250-V peak signals, dependent on the country or environment where the equipment is employed. The sinusoid associated with the ringing will ideally be generated locally, thus obviating the need for defining an analog *ringing bus* throughout the equipment (see Fig. 12-4). In this case the chosen semiconductor technology must be able to economically realize the logic circuitry which is needed to implement the sinusoid algorithms. An MOS technology

---

[3] The acronym BORSHT is attributed to Bell Laboratories, and it is frequently written in the German form as BORSCHT. In this case a codec is assumed to be included.

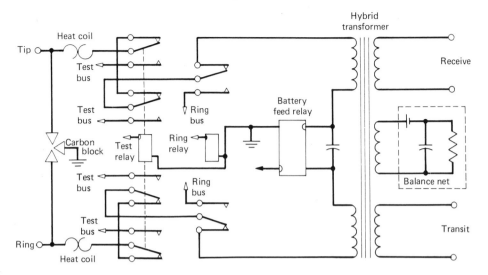

Tip-ring voice-frequency pair-cable connection

**Fig. 12-4** SLIC—traditional arrangement.

would appear to be the most likely candidate to satisfy the logic requirements, while an advanced linear technology will be needed to handle the high voltages involved. Several semiconductor manufacturers are busy pioneering such mixed technologies, and we can expect single-chip solutions to the SLIC in the foreseeable future.

## 12.3 DIGITAL SWITCHING

A digital switching machine may be represented by the blocks shown in Fig. 12-5. The individual analog lines enter the switch via a concentrator, or multiplexer module, which surrounds the core of the machine as shown. The concentrators digitize the message by passing the analog signal through the SLIC, filter, and PCM codec, and then combine the serial systems into a multiplexed form (see Fig. 12-1).

Some PABX and central office systems perform the concentration using the standardized 24- or 30-channel multiplex formats, and consequently include the frame formating (frame-code) information within the multiplexed digit stream. Other approaches take groups of 60 channels or more, while others use the 30-channel arrangement but do not insert the frame-code information. In this case frame synchronism is maintained by sending frame-timing command signals from the switch to the dependent concentrators.

Irrespective of which system is used, the end result is the same: several digit streams containing word-interleaved pieces of channel messages arrive at the core of the switching unit. This is shown in Fig. 12-5, where subscriber $S1/A$ wishes to be connected to $S8/G$, and $S1/D$ to $S4/F$. In the example, the digital words related to subscriber $S1/A$ are deposited (written) into a semiconductor memory, or store, at location $L1/A$, which is at the same level as the concentrator supplying the input information. Similarly, digital words relating to $S1/D$ are deposited within memory location $L1/D$.

We must now move the information word from $L1/A$ to $L8/G$. This implies a

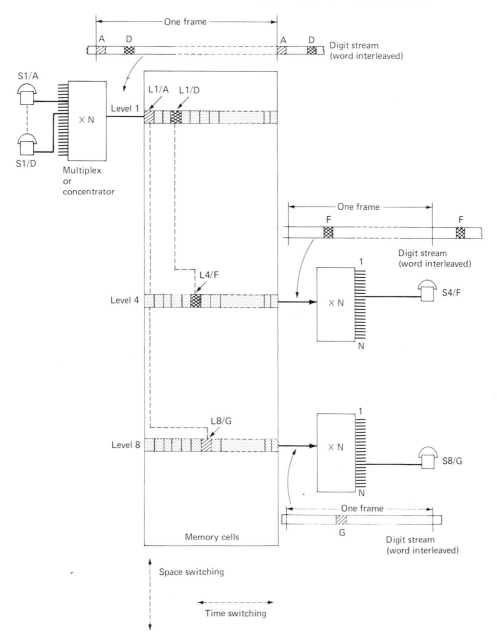

**Fig. 12-5** Space- and time-switching digitally encoded signals.

change in level $L1$ to $L8$, and is called *space switching*. Additionally, a change in time is also involved moving from position $A$ to $G$, and this is referred to as *time switching*.

During the reading operation the digital words, now at the new location, are formatted once more into a concentrated digit stream. The deconcentrator, or demultiplexer, allocates the message to subscriber $S4/F$ and $S8/G$ as shown.

The space-switching task can be simply implemented by connecting level 1 to levels

**Table 12-3  Percentage Distribution of Costs for
Analog- and Digital-Switching Systems**

| Cost contributor | Switching system type | |
|---|---|---|
| | Analog | Digital |
| Channel-level circuitry | 25 | 55 |
| Control circuitry | 25 | 30 |
| Network installation | 50 | 25 |

4 and 8 during the appropriate time slots.  Time switching, on the other hand, is carried out by manipulation of the memory cells relative to the time domain.

In theory it would be possible to build a switch based exclusively on time-switching, or alternatively space-switching principles.  In practice, however, the switching speeds needed for a system based entirely on time-switching principles would be prohibitive and consequently some form of space switching becomes necessary.

It is clear that the switching operation itself is highly dependent on the availability of low-cost fast memories.  Most systems in production today have standardized on the 16-kbit dynamic random access memory, however, we can anticipate moves toward 64-kbit and 256-kbit types as these become readily available at low cost during the eighties.

The actual realization of space and time division switching can take many forms. The above description has dealt with one method that is conceptually easy to understand and is memory intensive.

Space switching may be economically implemented using a hard-wired gated digital bus.  Let us return to the original example, where the information word $L1/A$ should be switched to $L8/G$.  The space-switching problem ($L1$ to $L8$) could alternatively be implemented by hard-wire connecting levels 1 and 8 during the time slot allocated to channel $A$.  This approach obviously reduces the amount of costly memories consumed, however, the read/write speed needed for the remaining time-slot interchange memories is likely to increase.  This must be true since the number of channels on a bus dictates the read-write speed requirement.  The upper limit set by the fastest technologies available (emitter, coupled logic) would appear to be in the region of 1000 channels maximum per bus.[4]

An additional complication that makes switching equipments currently difficult to classify is that several designs utilize both analog- and digital-switching techniques within the same equipment.  The Western Electric no. 101 ESS, for example, uses time division switching of analog TDM-PAM signals.  The actual switching element used is a normal field-effect transistor (FET) analog gate.  Several other equipments (e.g., E10 in France) use analog concentration at the first switching level, and then employ digital-switching techniques at the higher concentration levels.

The economic advantages of digital switching compared to analog systems relate to both initial equipment costs and overall network installation costs.  An approximate distribution of the costs involved for both analog and digital systems is shown in Table 12-3.  Although the channel-level circuitry is currently much higher for digital systems compared to analog, these differences are likely to disappear rapidly as the SLIC, codec, filter, and channel-level devices become low-cost commodity products.

[4] G. Perucca, "An Experimental Digital Switching System," *Proceedings of the International Switching Symposium,* Munich, Germany, Sept. 9–13, 1974, paper 227, 8 pp.

Similarly, we can expect continued cost reductions for memory devices as the cost per bit continues to drop (see Fig. 12-6).

### 12.3.1 Stored Program Control

Control of the switching operation described above is maintained by computer systems that interpret the dial (signaling) pulses, and allocate memory cells accordingly. Traditionally, the control has been an integral part of the switching apparatus. The electromechanical, step-by-step (Strowger) systems, for example, effectively performed both the switching function and the logical interpretation of the dial pulses within one operation. Some of the early system limitations were overcome using a *common control* approach. In this way the dialed pulses were stored separately, and could be interpreted using both electronic and electromechanical elements.

Stored program control (SPC) was a real breakthrough in switching control technology. SPC is in effect a separate subsystem within the switching machine that controls, using a given set of instructions, the switching operation. Since the instructions may be changed by a simple reprogramming of the machine, maximum flexibility can be achieved. In the early days, due to the higher start-up cost for a common control system, the approach could only be used in large switching machines. The availability of low-cost powerful 16-bit microprocessors has changed this, and now SPC is finding a place in the small PABX market also. In essence microprocessor technology has permitted:

1. Economic implementation of SPC within a small switching system
2. Implementation of SPC within a large system, using distributed control in a multiprocessor configuration, as opposed to a centralized monoprocessor

It is pertinent to study what SPC really does give in benefits to the subscriber and to the operating post, telephone, and telegraph (PTT). Let us take the example of how a *distribution frame* can be removed by software programming as a case in point.

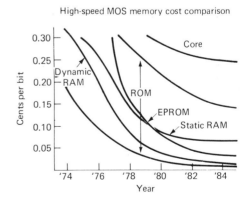

**Fig. 12-6** Projected costs per bit for different memory devices. (From M. Shepherd, "The Promise of the Microelectronic Revolution," unpublished conference paper presented by the chairman of Texas Instruments at the Institute of Technology, Chicago, Oct. 3, 1979.)

The need for distribution frames that physically patch wires between the switching machine outlet and the subscriber pair can now be dispensed with (see Chap. 2). The SPC maintains a permanent register of subscriber destinations versus the assigned telephone number, which can be updated as necessary. These registers will in the very near future be contained on nonvolatile magnetic bubble memories containing 1 million bits or more.

It is clear that the computer control system is of major importance to the total operation of the digital switch. However, it should not be overlooked that many system features can easily be added by software changes within the computer programs. Within PABX systems, the inclusion of special features is of particular importance. The following list is typical of the features that can now be offered:

1. *Maintenance:* Each of the memory cells can be routinely checked when out of traffic.
2. *Call charge:* The total cost and individual cost of each call can be separately listed together with the dialed number for each subscriber. This is particularly useful in factory PABX installations.
3. *Long-distance number blocking:* Each subscriber can be simply blocked from making calls outside the PABX: a long distance call, or an international call, as required.
4. *Follow-me:* By temporarily changing the telephone number assignment register, calls can be transferred from one location to another automatically. A typical example would be secretarial transfer.
5. *Don't disturb:* Any phone can be temporarily blocked against incoming calls, and calls rerouted if required.

Obviously the list is almost endless, as a visit to any modern PABX will show. The key is software programming, and this gave to a number of computer manufacturers a competitive edge against those telecommunication manufacturers who continued during the seventies to market traditional switching solutions at similar cost. In the eighties we may expect more standardization of software languages, such as CCITT's High Intelligence Level Language (CHILL), which is specifically tailored for telecommunication switching.

### 12.3.2  Local Concentrators

It is clear, by analysis of Fig. 12-5, that each subscriber must be linked by a unique pair of wires to the exchange. In recent years several attempts have been made to better utilize the local cable network by introducing PCM multiplexing systems nearer to the subset itself. We can foresee, therefore, a gradual move from today's approach to digital switching, typified by Fig. 12-7a, toward the schemes shown in Figs. 12-7b, c, and d.

Offices and hotels or wherever there is a high concentration of telephone traffic, plus a local power supply, favors the approach shown in Fig. 12-7b. In such systems, which have already been put into service in North America and Europe, standard 24- or 30-channel multiplexers are used. The digitized multiplexed signals are passed directly to the time-slot interchange memories of the switching machine. Thus, the need for two coding steps is avoided. This type of equipment is known as *subscriber PCM*.

Fig. 12-7c shows a similar approach, although fewer channels are multiplexed. These are called *local PCM* systems. The number of multiplexed channels has as

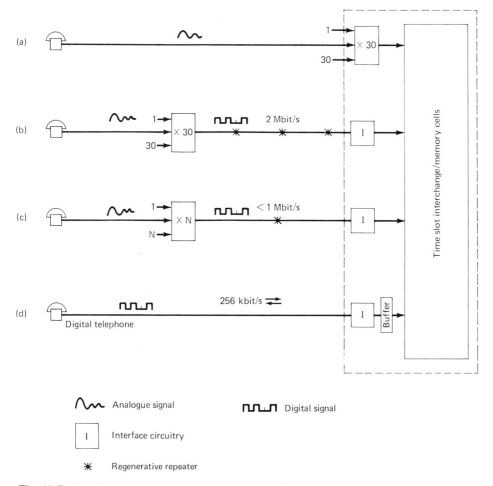

m̃  Analogue signal          ⊓⎍⊓  Digital signal

| I |  Interface circuitry

✳  Regenerative repeater

**Fig. 12-7**  Local concentration of subscriber signals: (a) conventional analog subscriber connection; (b) concentration using a subscriber PCM equipment (24 or 30 channel); (c) concentration using a local PCM equipment (6, 8, or 10 channel); (d) digital subset without concentration (see Fig. 12-8).

yet to be agreed upon internationally, although it is already clear that there will be far fewer than the 24 or 30 channels normally combined.  (The United Kingdom, Italy, and Norway have proposed 6, 8, 10 channels.)  The attraction of local PCM systems compared to the subscriber PCM approach is the low transmission bit rate that can be used.  The consequential reduction in attenuation and near-end crosstalk in particular permits far greater repeater spacings and hence cost savings.  The eventual limit in the digitization of the telephone network will have been reached when the codec becomes part of the telephone subset itself.  This case is considered in Fig. 12-7*d*.

## 12.4  DIGITAL TELEPHONE SUBSETS

The digitization of the voice signal within the telephone subset will be as revolutionary to the telecommunication industry as digital switching.  The introduction of a codec within the subset, and the implied digital transmission to the switching machine is in itself not very startling. It is true, this will reduce the overall system cost, the size of the terminal equipments, and provide a higher quality of speech communication to the subscriber. However, the real importance of digital communication directly to the subset will be seen as society demands more and more access to data retrieval systems, and facsimile machines move into general usage within the home, warehouse, or office. These home information systems can be provided at very low cost, without

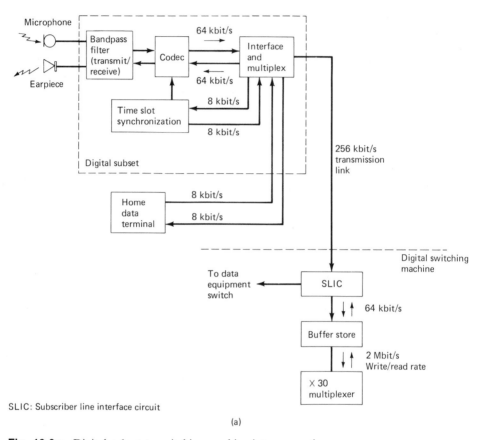

SLIC: Subscriber line interface circuit

(a)

**Fig. 12-8a**    Digital subset to switching machine interconnection.

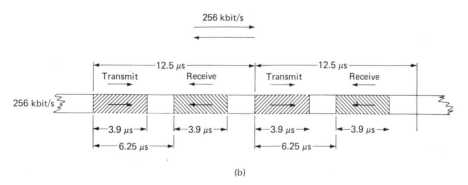

(b)

**Fig. 12-8b**    Bidirectional transmission over a time-shared link.

the need of modems, and may operate simultaneously with voice messages provided that the speech is first digitized. Imagine the ability to discuss on the telephone while *simultaneously* making drawings using a facsimile machine, without any provision for additional transmission equipment!

As a pointer to the future, it is pertinent to describe the field trial digital telephone introduced by ITT-KIRK within the Jutland telephone network in Denmark. This system, believed to be the first in the world to include a codec in the subset, began field trials during 1979. A block schematic of this system is illustrated in Fig. 12-8. The information rate required in each direction for this system is 80 kbit/s compiled as follows:

1. Codec voice information (8 kHz sampling, 8-bit words): 64 kbit/s
2. Data channel, (available for facsimile, etc.): 8 kbit/s
3. Signaling synchronization, and supervision data: 8 kbit/s

Bidirectional communication is obtained using a single subscriber pair by time sharing within a 256 kbit/s transmission scheme, as shown in Fig. 12-8b. Thus the transmit direction is allocated the first 3.9 $\mu$s ($\frac{1}{256}$ kbit/s) of every 12.5-$\mu$s ($\frac{1}{80}$ kbit/s) time slot; while the receive direction is constrained to lie within the second half of the same time period. A guard band of 4.7 $\mu$s is used to avoid aliasing errors. The system uses a low-power CMOS codec which need only be written or read at a maximum rate of 64 kbit/s, rather than the conventional 1.544 or 2.048 Mbit/s interrogation speeds. Thus the power consumption and constraints imposed on the codec design can be further reduced. The interface requirements at the switching machine must address the following problems:

1. Power feed to the subscriber subset when the subscriber line is in use only
2. Bit rate conversion from 64 kbit/s to 1.544 or 2.048 Mbit/s interrogation rates (necessitates a buffer store)
3. Separation of signaling and supervision data
4. Synchronization of the channel

It is pertinent to describe the synchronism problem in more detail. Clock synchronism difficulties between the transmit and receive directions are most easily overcome by using the same reference clock. The reference clock is provided by the switch and is extracted at the subset, where it is used for transmission back to the switch, plus sampling of the message signal, and so on. Frame synchronism is also referenced to the switch, which provides framing information to define the start of each 12.5-$\mu$s time-slot period. A block diagram of the telephone subset circuitry is shown in Fig. 12-8a.

## 12.5    OPTICAL COMMUNICATIONS

Most of the properties of guiding optical rays along a fiber can be described, at least qualitatively, in terms of simple geometrical optics. If an optical ray is incident on a boundary between two optically transparent media, yet of different refractive index, then part of the light is refracted and part is totally internally reflected (see Fig. 12-9). The refraction process is defined by Snell's law as

$$n_1 \sin \theta_1 = n_2 \sin \theta_2 \qquad (12.1)$$

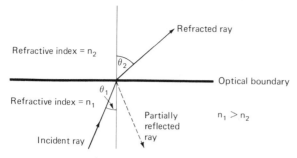

**Fig. 12-9**  Optical refraction.

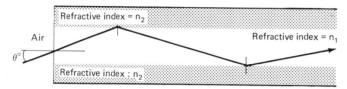

**Fig. 12-10**  Multiple internal reflections within a fiber.

where $n_1$ = refractive index medium 1

$n_2$ = refractive index medium 2

If the second medium has a lower refractive index than the first (i.e., $n_1 > n_2$) there is a critical angle where the above equation cannot be satisfied. This occurs when

$$\sin \theta_2 = \frac{n_1}{n_2} \sin \theta_1 \geq 1 \qquad (12.2)$$

In this case no refraction occurs, and the incident ray is totally internally reflected. The case of multiple internal reflections is shown in Fig. 12-10, and this is essentially what happens in an optical fiber, although here the situation is somewhat more complex due to the curvature of the optical boundaries. Snell's law may be extended to calculate the maximum angle $\theta_m$ that a ray may enter a fiber and undergo guided propagation:

$$\sin \theta_m = \sqrt{(n_1{}^2 - n_2{}^2)} \qquad (12.3)$$

This quantity is called the *numerical aperture* of the fiber. Rays incident at larger angles are only partially reflected, on each occasion, and are very rapidly lost.

The entry angle $\theta$ of the optical ray can support discrete values only; each particular value being identified as a *mode*. The total number of modes supported by a fiber is dependent on both the core area and the numerical aperture of the fiber as given by:[5]

$$\text{Number of modules} \stackrel{\sim}{=} \frac{2\pi}{\lambda^2} \times \text{core area} \times (\text{numerical aperture})^2 \qquad (12.4)$$

[5] C. P. Sandbank, "Fiber Optic Communications: A Survey," *ITT Electrical Communications*, vol. 50, no. 1, 1975, pp. 20–27.

Clearly, it is possible to restrict the number of modes that can propagate along the fiber by reducing the *core area,* or numerical aperture, or both. In the limit the fiber can be restricted to permit the propagation of one mode only, and is referred to as a *single-mode fiber,* as opposed to a *multimode fiber* type.

The system benefits gained by the usage of single-mode fibers are related mainly to the low pulse-dispersion characteristics of such a medium. In an ideal system each pulse of light, represented by several optical rays, will travel along the fiber and arrive simultaneously at the receiving end. In practice, however, multimode fibers permit some rays to travel the most direct route ($\theta = 90°$), while others may take a much longer path, undergoing many reflections on the way. In this case the pulse seen at the receiving end of fiber will be broadened (i.e., dispersed) as contributions arrive at different times from the various optical rays. In the limit, successive pulses will completely overlap each other, and make detection impossible (see Sec. 3.7).

Let us consider the pulse broadening $\Delta t$ due to a fiber of length $L$. We need to compare the optical path lengths for a direct ray ($\theta = 90°$), and an extreme ray ($\theta = $ critical angle). Therefore,

$$\Delta t = \frac{L}{C}(n_2 - n_1) \tag{12.5}$$

where $C = $ speed of light
$n_1, n_2 = $ refractive indices of the glass media

Typically, we may suppose a 1 percent difference in the refractive indices of the glass, about a value of 1.5. In this case, using the above formula for a value of $L$ equal to 2 km, we obtain a pulse broadening of order 100 ns. Thus, the maximum usable bandwidth $B$ of such a link would be of order 10 MHz.

The above analysis gives a somewhat pessimistic result compared to experimental evidence. The approach has assumed that the product of path length $L$ and bandwith $B$ is a constant. This is not true for long fiber lengths, where an inverse-root path-length dependence is seen. Consequently we may expect to see a maximum usable data rate in the region of 80 to 100 Mhz in a practical system.

The foregoing would suggest that single-mode fibers should be the natural choice for all systems employing a high data rate. This is essentially true provided an optical source of very narrow spectral distribution is chosen. If this criterion is not met, then *material dispersion* will occur. This phenomenon is due to the fact that different wavelength components travel at different velocities within an optical medium. The effect is an order of magnitude lower than the mode dispersion described above, as shown in Table 12-4. Consequently material dispersion becomes a limiting factor for laser sources over distances greater than 5 km at bit rates of order 1 gigabit per second (Gbit/s) or more.

**Table 12-4   Material Dispersion Due to Different Optical Sources**

| Optical source used over 2 km glass fiber | Typical wavelength spread, nm | Pulse broadening $\Delta t$, ns |
|---|---|---|
| Light emitting diode | 20.0 | 3.60 |
| Gallium arsenide laser | 2.0 | 0.36 |
| Neodymium doped YAG laser | 0.2 | 0.04 |

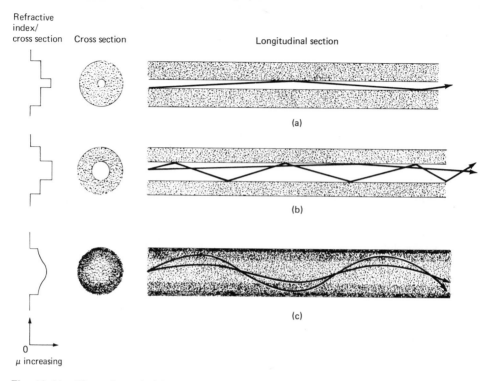

Refractive index/cross section    Cross section    Longitudinal section

(a)

(b)

(c)

0
μ increasing

**Fig. 12-11**   The main optical fiber types: (a) single-mode fiber; (b) multimode fiber; (c) graded-index fiber.

An alternative approach to overcoming the dispersion problem is to use a *graded-index* fiber.   In this case the refractive index does not vary as a step-function between the two media, but gradually decreases in value away from the fiber axis (see Fig. 12-11*c*).   Fibers of this construction have the property of continuously refocusing the optical ray as it passes along the cable, thus minimizing the effect of dispersion. In terms of geometrical optics, the guided rays no longer follow zigzag paths down the fiber, but track a smooth, almost sinusoidal path.

Mode dispersion is significantly reduced in a graded-index fiber since the lower-order modes that travel almost axially are forced to propagate through the high-refractive-index region of the fiber.   Consequently the velocity of such axial rays is reduced compared to the higher-order modes, which tend to travel through the low-refractive-index region.   Thus the path delay between the various optical modes is somewhat compensated by promoting the relatively faster propagation of the higher-order modes through the fiber.

Graded-index fibers were first introduced in Japan by the Nippon Electric Co. (NEC) under the trademark Selfoc (self-focusing).   They typically have a low attenuation, and are well suited to high data-rate communication systems.   The main drawback at the present time remains the high cost of such cables compared to conventional copper.

It is pertinent to list the advantages of optical communication compared to conventional copper cable:

### 1. SMALL PHYSICAL SIZE

Permits the introduction of additional circuits within overcrowded city ducts.

In many cases today's high optical cable costs can be offset against the cost of building new ducts across cities.

## 2. LOW WEIGHT
Important in fuel-economic cars, and airplanes. Also a benefit during installation of telephony cables.

## 3. IMMUNE TO ELECTRICAL INTERFERENCE
Particularly important within industrial environments, and for short-distance interconnects in electrically noisy surroundings (e.g., switching exchanges).

## 4. OVERCOME GROUND-LOOP PROBLEMS
Beneficial for medium-length interconnects in oil tankers, airplanes, etc.

## 5. RADIATION RESISTANT
For military applications, and for use in nuclear power stations, etc.

## 6. PROTECTION AGAINST OVERVOLTAGE SURGES
The insulation properties of glass obviate the need for protection circuitry against lightning discharge, etc.

## 7. SECURITY
It is very difficult, if not impossible, to intercept a communication via optical fiber.

## 8. FEWER REPEATERS
Present-day technology has produced optical cables of less than 1-dB attenuation per kilometer (dB/km) at high bit rates. Fig. 12-12 would suggest the

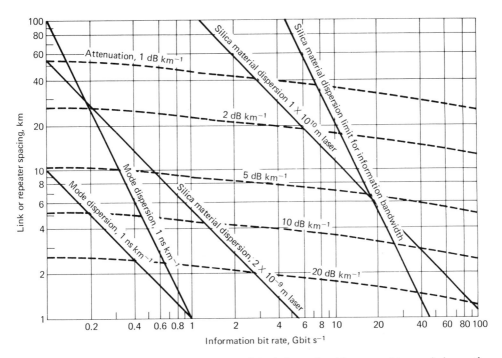

**Fig. 12-12** Repeater spacing as a function of the information bit rate and transmission path attenuation. (From M. M. Ramsay, G. A. Hockham, and C. Kao, "Propagation in Optical Fiber Waveguides," *ITT Electrical Communications,* vol. 50, no. 3, 1975.)

feasibility of repeater spacings in the region of 50 km, at data rates of 200 Mbit/s. However, the real benefits will be seen at data rates of order 6 to 40 Mbit/s where no repeaters will be required on typical junction routes.

9. HIGHLY FLEXIBLE

A sheathed optical fiber cable remains extremely flexible, and much easier to lay compared to conventional copper.

10. LOW COST

Optical fiber costs have reduced dramatically in recent years as technology and production volume have increased. The basic raw materials such as glass, or silica (quartz) are readily available, and in any case only small quantities are used within each fiber. Cost comparisons between optical communication and copper-based systems over long distances are already tending to favor the optical approach. For short interconnects the cost comparison between two systems comprising line drivers and receivers (optical, or electrical current) is promoting the rapid introduction of optical communication within noisy environments.

### 12.5.1  Optimization of the Optical Transmission Characteristics

Each optical medium exhibits a particular absorption spectra (see Fig. 12-13) dependent on the impurity ions within the material. The spectral peak at 945 nanometers (nm) is due to the hydroxyl (OH) ion that can occur within glass, and must be removed during processing to ensure that the completed fiber has low attenuation. It is this constraint that has led to the need for processing technologies similar to those used in the semiconductor industry. Fibers with attenuations of less than 2 dB/km, and negligible peaks at 945 nm, have been produced indicating that it is

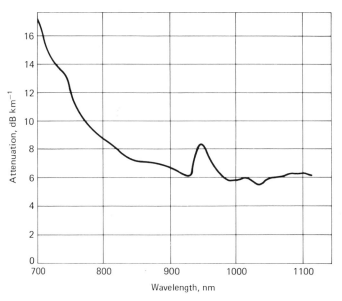

**Fig. 12-13**  Typical attenuation within a germanium oxide–doped silica fiber over 1 km in length. (From C. P. Sandbank, "Fiber Optic Communications:  A Survey," *ITT Electrical Communications,* vol. 50, no. 1, 1975.)

feasible to build fibers with attenuations approaching that due entirely to Rayleigh scattering.

The Rayleigh scattering loss may be attributed to irregularities inherent in the glassy state, and is related to the wavelength of the optical source as:

$$\text{Rayleigh scattering loss, dB} \propto \frac{1}{(\text{source wavelength})^4} \qquad (12.6)$$

Thus, there is a very real advantage in choosing optical emitters that operate in the infrared region as opposed to the visible region of the spectrum. Fortunately, optical emitters and detectors are readily available within the required wavelength range: for example, light-emitting diodes (850 to 1200 nm), solid-state lasers (1000 nm), silicon pin diodes, and avalanche photo-diode detectors. [Considerable work is underway to produce higher wavelength light-emitting diodes (LEDs), since fiber losses are lower in this region.]

The beam emitted by a laser is extremely narrow, and permits a highly efficient launching of the optical ray into the fiber. Thus fibers of small numerical aperture, and hence low dispersion, may be used. By contrast LEDs emit rays over a wide field, and are difficult to couple into fibers.

The energy launched into a fiber is related to the number of modes transmitted. Consequently the energy launched is proportional to both the core area and the numerical aperture squared [Eq. (12.4)]. Thus, for low data-rate systems, where dispersion is less important, large numerical aperture fibers are preferred. LEDs are typically used in such systems, since these can be effectively coupled into the larger core areas employed.

The trade-off between LED and laser emitters is not obvious. Each system will have different requirements dependent on data rate and transmission distance. For short-haul, low data-rate systems, the low-cost LED has definite advantages. On the other hand, the efficient launching characteristics and low material dispersion associated with laser emitters will favor their usage in high-capacity systems.

The first systems employing laser emitters were notoriously unreliable. These problems appear now to be solved, with lifetimes of over 10,000 h already recorded. Such results have been obtained by removing the failure mechanisms within the device itself, and by limiting the duty factor to less than 30 percent typically.

We shall now consider some of the specific system problems associated with optical communication.

### 12.5.2  Optical Repeater

The main difference between an optical regenerative repeater and a conventional type lies in the equalization circuitry (see Fig. 12-14). Traditional equalization mechanisms compensate for the nonlinear frequency attenuation characteristic of conductive cables, and perform some pulse shaping (see Chap. 11). In an optical system the attenuation is completely unrelated to the data transmission rate, and is a function of the optical absorption at the wavelength used by the optical emitter, plus the effects of Rayleigh scattering.

The equalizers used in optical systems are designed to compensate for the effects of dispersion. In many systems where the dispersion is small compared to the bit period no equalization correction is required. However, in some high-capacity systems this happy state of affairs does not exist, and dispersion equalization is employed.

The ideal equalizer will produce an output peak at time $t = 0$, and pass through

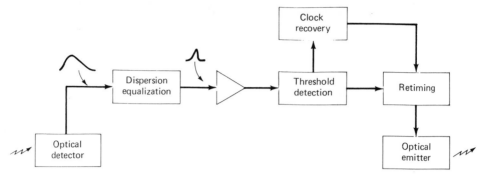

**Fig. 12-14**  Block diagram of an optical regenerative repeater.

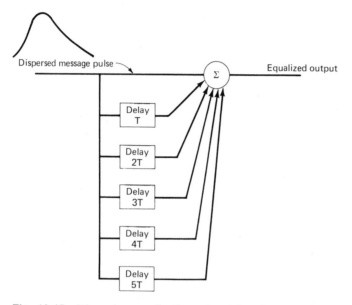

**Fig. 12-15**  Dispersion equalization using delay elements.

zero at all instants $t = nT$ where $T$ represents the bit period.  In this case the original message can be reconstructed with zero intersymbol interference by sampling the equalized signal at instants $t = nT$.

The implementation of a dispersion equalizer can take the form of an *RLC* filter, or a transversal filter, or a combination of the two.  The transversal filter approach is shown in Fig. 12-15, where the equalized output signal is obtained by cross-correlation of pulses at $t = 1, 2, 3$, and so on.  It should be noted, however, that this technique is not suitable on its own when the dispersion has a long tail, since a large number of delay elements would be required.

The degree of dispersion correction required will vary between fiber samples, and the repeater separation.  Consequently an adaptive approach to equalization is desirable.  One approach to adaptive dispersion equalization that has been suggested[6] uses

[6] M. M. Ramsay, A. W. Horsley, and R. E. Epworth, "Subsystems for Optical-Fiber-Communication Field Demonstrations," *Proceedings of the IEE,* vol. 123, June 1976, pp. 633–642.

the comparison between the frequency spectrum of a dispersed and nondispersed random test signal as a measure of the needed correction. The nondispersed spectrum is that obtained for a fiber of zero length and can be predicted theoretically. The comparison between the two spectra need be made at only one or two selected frequencies, rather than across the entire range. Thus the implementation of this approach is not too complex.

The remaining circuit blocks within an optical regenerative repeater are identified within Fig. 12-14. Since their function is essentially equivalent to a conventional repeater (see Chap. 8), no further description is required.

### 12.5.3 Coding within an Optical System

We have already analyzed the reasons for coding a message signal prior to its transmission (Chaps. 8 and 9). The methods employed in conventional conductive cable systems are based on ternary (multilevel) transmission, and are consequently not easily applicable to the optical case. Most optical systems employ direct modulation of the optical emitter. That is, electrical binary 1s and 0s are converted to flashes of light which are either present or absent; there is no third variable. (Phase modulation of the optical carrier, or transmission at different optical wavelengths, is theoretically possible but difficult to implement.)

The first optical systems used redundant binary block codes such as 5B6B. In this example the coder is arranged to convert groups, or blocks of 5 binary digits into a group of 6 binary digits which are represented optically. The technique is similar to 4B3T coding (see Sec. 9.4.4), however, in the case of 5B6B the transmission bit rate actually increases. The extra redundancy can be very small, however, the penalty of higher circuit complexity for the coding, decoding, and error detection circuitry becomes significant, and a limiting factor, as the block length increases.

Another approach is to use a *scrambled binary* coding scheme. This has all the required properties with the exception of error monitoring, which can be provided by the insertion of a parity bit. The implementation of such a scheme is considerably more economic in terms of circuitry compared to the block coding methods.

# 13

# Digital Recording

## 13.1 INTRODUCTION

The similarity between the techniques involved in digital communications and recording have been referred to throughout this book. It is appropriate, therefore, to devote this final chapter to the particular problems of digital recording. We shall commence this analysis by trying to better understand the motivation for adopting digital techniques within the recording process, and follow this by a detailed study of the benefits digitization can provide within the domestic environment.

There must have been many technological innovations in recent years within the hi-fi music field that have improved sound quality, albeit slightly, and have lightened pocketbooks considerably. It would be excusable therefore to dismiss digital recording technology as yet another minuscule delta improvement, compared to a sound quality from current state-of-the-art analog systems which most people already find quite satisfactory. The fully digital disc is likely to change this perception of digitization within the mind of the general public.

Let us digress for a moment and compare the evolution of hi-fi sound quality to the development of photography. In the early days of photography a group of serious amateurs and professionals known as the *circle of confusion*[1] used to meet in order to exchange ideas and information about their hobby in general. On one occasion, a member of the group showed a print that had been enlarged to 11 × 14 inches (in) from a 35-mm negative. "Look, no grain!" he said. "Then we took a fraction of that negative, and blew that up to 11 × 14 inches. See, no grain!" Another member of the group, A. Boni, said, "You guys are all whacky, you are going just the other way!" When asked for an explanation he declined, and walked off to build a million-dollar business known as *Readex Microprint,* where large documents are photographically reduced in size!

The relevance of the above story to this introduction of the benefits offered by digital recording is that most analyses of this subject to date have concerned themselves with improvements in the studio process of producing an analog disc. These improvements are undoubtedly worthwhile and important, however, if we take a longer-term view the domestic replay of digital discs could make records smaller, more manageable, and able to be handled even with dirty hands. Combine those advantages

---

[1] "How the Leica Changed Photography," in Douglas O. Morgan and David Vestal (eds.), *Leica Manual: The Complete Book of 35mm Photography,* 15th ed., Morgan, Dobbs Ferry, New York, 1973.

with the elimination of static pickup, print-through[2] problems in the case of magnetic tape systems, and zero wow and flutter,[3] and it is clear that there are far more benefits to be obtained by digitization than the obvious signal-to-noise ratio improvement when compared to analog systems. Today, as the photographic hobbyists did in the sixties, we need to take a wider look at the implications of digital recording.

We shall now examine the main processes involved in the production and reproduction of today's analog phonograph record.

## 13.2    THE PRODUCTION AND DOMESTIC REPLAY OF AN ANALOG PHONOGRAPH RECORD

The process of storing the sounds derived from a live performance on a phonographic record, and replaying them faithfully within a domestic environment, is depicted simplistically in Fig. 13-1. There are three main processes involved:

1. Master recording
2. Record manufacture
3. Domestic reproduction

It is instructive to examine the various processes imposed on the original acoustical signal as it is converted from a pressure wave to an electrical voltage (microphone), to a magnetic pattern on a tape (recording), to a positional variation in the groove walls on the record (disc cutting), and back again to yield the "original" acoustical signal. Each of these conversion steps contributes a small, but noticeable impairment on the fidelity of the final reproduction.

Today's best microphones, as typically used for professional recordings, produce electrical voltage variations that conform very precisely with the air pressure changes that they represent. Similarly, the master disc-cutting, electroplating, and pressing processes are remarkably free from distortion provided due skill and care is exercised by the cutting engineer, and the cutting equipment is set up accurately. On the other hand, the master tape recorder can introduce serious audible distortions which become more noticeable if the original master tape is recorded several times.

Some specialized companies have attempted to eliminate the degradations introduced by the master tape recorder by driving the master lacquer disc-cutting machine directly from signals obtained from the mixing console (see Fig. 13-1). These records, known as *direct-cut,* require that each record side must be recorded in a single session, and pose severe restrictions from an artistic point of view. Errors cannot be corrected, or edited out, and as a consequence any mistake necessitates a retake of the complete side. For these and other reasons, recording engineers have concentrated their efforts at improving the quality of the master tape recording process itself.

Analog tape recorders introduce two major sources of degradation. The first is caused by the inherent lack of linearity between the analog message signal and its magnetized representation on the tape. The second is due to the active coating on the recording tape, which is composed of very small magnetic particles held together by a nonmagnetic binding material. In the limit, the size of these particles introduce a noise background, or a continuous hiss, and the purity of the reproduced musical note is degraded.

---

[2] Print-through: The phenomenon of recorded signals from adjacent layers of a recording tape interfering with each other.

[3] Wow and flutter: Distortion effect due to momentary variations in the playback speed of the reproduction equipment. This causes variations of pitch within the musical signal.

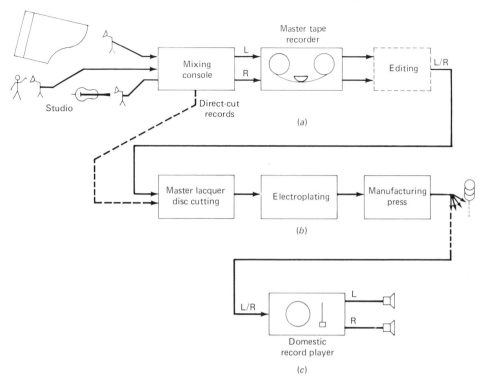

**Fig. 13-1** A simplified representation of the production and domestic replay of an analog phonograph record: (a) master recording; (b) record manufacture; (c) domestic reproduction.

It has already been pointed out that the original multitrack master tape is normally edited many times, and additional stages of mixing are required before a final stereo (dual track) master recording is produced. During these rerecording sessions one can expect a gradual buildup of the degradation described above. Furthermore, the effect of occasional tape drop-outs, due to missing magnetic particles on the tape, will become more noticeable.

Finally, variations in the tape playback speed, due to slippages between the tape recorder capstan and flywheel, will introduce a change of pitch within the reproduced music signal. This effect is known as *wow and flutter*.

The attraction of digital recording is that since the tape recorder need only identify the presence or absence of digital pulses, the signal-to-noise and lack-of-linearity problems associated with analog signals are completely eliminated. The introduction of an error-correcting coding structure can be used to overcome the momentary loss of signal associated with drop-outs. Finally, by retiming the retrieved digital signal using a stable and accurate timing source, such as a quartz crystal, it is possible to correct for wow and flutter.

## 13.3   A COMPARISON BETWEEN ANALOG AND DIGITAL RECORDING

The conventional practice[4] in professional recording is to use a 16-bit uniformly quantized (i.e., no companding) system, sampled at 48 kHz. In this case the maximum theoretical dynamic range is given by:

---

[4] John Atkinson, "Practical Digital Recording," *HI-FI News and Record Review Magazine,* August 1979, pp. 69–73.

$$\text{Dynamic range} = 20 \log_{10} \left( \frac{\text{largest signal value}}{\text{smallest signal value}} \right) \quad \text{dB} \qquad (13.1)$$

$$\text{Dynamic range} = 20 \log_{10} \left( \frac{2^{16} - 1}{1} \right) = 96.33 \text{ dB} \qquad (13.2)$$

In practice a dynamic range of order 90 dB is typical since the least significant bit is never very accurate, and some degradation in performance usually occurs within the analog input circuitry.[5] By contrast the best analog systems provide a dynamic range of order 72 dB at 3 percent distortion.

The main benefit of digital recording is that very low levels of distortion can be realized compared to analog systems. Theoretically the harmonic distortion of a digital system is zero. However, due to practical considerations a value of 0.03 percent is typical, which compares favorably to that of analog equipments which lie in the 1 to 3 percent range. Similarly, the intermodulation distortion of a practical digital system is of order 0.15 percent, compared to a typical value of 10 percent for analog equipments. We can expect even lower values of distortion in future years as integrated circuits having more accurately defined linear $A/D$ and $D/A$ tracking parameters become available. However, the audible benefits will be minimal unless similar improvements occur in loudspeaker design, etc. There are few loudspeakers that can handle loud signals at 0.1 percent distortion.

The $A/D$ and $D/A$ converters used within 16-bit professional systems are extremely complex circuits, and force the usage of considerable quantities of recording tape compared to, say, a simple 14-bit structure. It is therefore reasonable to inquire, What is the optimum word length for digital recording systems? The British Broadcasting Corporation (BBC) has demonstrated that the difference between the original and the retrieved digital signal is almost imperceptible for *most* sound sources when the word length is 12 bits or more. In fact the BBC has successfully employed for many years a 13-bit system sampled at 32 kHz to interconnect their radio transmitters with the studio using standard telephone lines. The system provides a 70-dB dynamic range and a virtually flat frequency response up to about 14.5 kHz. It can be argued that due to advances in microphone quality within the studio, and improved domestic reproduction equipment, there is merit in extending the dynamic range compared to that achievable within a 13-bit structure. However, for most domestic purposes a 14-bit system would appear to provide the optimum performance-cost trade-off.

A 16-bit structure provides an additional 12 dB of range, which offers the recording engineer a margin of safety to avoid peak clipping problems. The audible effect of digital peak clipping is to introduce a very loud, undesirable "crack," which can be quite long in duration for severe overloading. This phenomenon effectively forces an engineer using digital recording equipment to restrict the normal working range to 14 bits in case the odd musical sound, or transient, exceeds the peak-to-peak value that can be accommodated by the most significant bit.

It should be noted that analog recording systems are more tolerant to peak overloads than digital recorders. The explanation of this is that analog machines tend to destroy the phase coherence of any transient sounds, so that the sudden impact of energy

---

[5] The signal-to-quantization-noise ratio ranges from a best case of 90 dB (maximum signal level) to zero (minimum signal level) using the same analysis as Eq. (13.1). This parameter is not very useful in comparing digital to analog systems where a third parameter, fidelity, or distortion is also important.

is diffused in time rather than acting instantaneously. By contrast, a digital recorder tends to track any transients perfectly, such that the peak-to-peak level is preserved within the audio range, as is the phase coherence. This causes two practical problems when using digital recording systems. First, as already mentioned a safety margin of order 10 dB must be preserved to avoid peak clipping. Second, the record groove velocities may become excessive and incapable of being tracked by even the best cartridges. For these reasons it is likely that professional digital recordings are likely to continue to be made using a 16-bit structure, and that these will be manipulated to yield a 14-bit word free of peak clipping for domestic usage. This manipulation can in practice be performed very easily by computers that can be programmed to avoid not only any peak clipping, but also limit the maximum permitted level for the record groove velocities, and even set a minimum groove velocity level to overcome turntable noises, known as *rumble*. It is a particularly attractive feature that the computer can interact directly on the digital signal, without the need to revert back to the analog format.

## 13.4 EXTENDING THE DIGITAL PROCESS

We have seen in Sec. 13.3 that digital recording techniques can produce useful improvements in the master recording process by eliminating the degradations produced by the analog tape recorder. In the medium term it is likely that the digital approach will be adopted within studio mixing consoles and electronic editing units. The latter could provide computer-assisted editing of digital data that is temporarily off-loaded into solid-state memory, where an inaudible, exactly executed, electronic edit is performed.

The above advances will yield significant improvements in audio quality and permit record manufacturers to build up a repertoire of high-quality master recordings. From the consumer's point of view he or she can continue to use a conventional analog record player without modification.

In the not so distant future it is already clear that fully digital systems will be available for domestic reproduction. This will undoubtedly further improve sound quality, but, more importantly, there will be a noticeable improvement in record handling. The domestic record player will need to be changed to reproduce the new fully digital records that are expected to be available soon. Already several manufacturers have exhibited prototype digital record systems that have been developed primarily with the video market in mind. Some of the systems, such as the ones pioneered by Philips, Eindhoven, use lasers and detect the reflected light which is modulated by the presence or absence of a pit on the surface of the disc. Other systems employ capacitive pickup from cells located spirally along the record surface. We can expect several years of field trials, market acceptance trials, and finally (hopefully!) international standardization before the digital disc becomes widely adopted.

# References

The publications and technical papers listed below will be found useful for supplementary reading and a more in-depth treatise of the described topics. Additional references will also be found as footnotes throughout the book.

## GENERAL OVERVIEW OF TECHNOLOGY

### Supplementary Reading

Bell Telephone Laboratories Technical Staff: *Transmission Systems for Communications,* 4th ed., rev., Bell Telephone Laboratories, Western Electric Co., Technical Publications, Winston-Salem, N.C., 1971.

Flood, J. E. (ed.): "Telecommunication Networks," *Institution of Electrical Engineers Telecommunication Series 1,* Peregrinus, Stevenage, England, 1977.

Hills, M. T.: *Telecommunications System Design,* vol. 1: *Transmission Systems,* vol. 2: *Switching Principles,* G. Allen, London, 1979.

Welch, S.: "Signalling in Telecommunication Networks," *Institution of Electrical Engineers Telecommunications Series 1,* Peregrinus, Stevenage, England, 1979.

### Background Reading

Aaron, M. R.: "PCM Transmission in the Exchange Plant," *Bell System Technical Journal,* vol. 41, 1962, p. 99.

Garratt, G. W. R.: "The Early History of Telegraphy," *Philips Technical Review,* vol. 26, 1965, pp. 265–268.

Halliwell, B. J.: "PCM and Digital Networks," in B. J. Halliwell, *Advanced Communication Systems,* Butterworths, London, 1974.

Hartley, R. V. L.: "Transmission of Information," *Bell System Technical Journal,* vol. 7, 1928, pp. 535–563.

————, P. Mornet, F. Ralph, and D. J. Tarran: *Techniques of Pulse Code Modulation in Communication Networks,* Cambridge University Press, London, 1967.

Martin, J.: "Digital Channels and PCM," in J. Martin, *Future Developments in Telecommunications,* 2d ed., Prentice-Hall, Englewood Cliffs, N.J., 1977.

## GENERAL BACKGROUND TO DIGITAL TELECOMMUNICATION TECHNIQUES

### Supplementary Reading

Lucky, R. W., J. Salz, and E. J. Weldon, Jr.: *Principles of Data Communication,* McGraw-Hill, New York, 1968.

## Specialized Papers (Highly Recommended)

Bose, R. C., and D. K. Ray-Chaudhuri: "Further Results on Error Correcting and Binary Group Codes," *IEEE Transactions on Information and Control,* vol. IC-3, 1960, p. 279.

Hamming, R. W.: "Error Detecting, and Error Correcting Codes," *Bell System Technical Journal,* vol. 29, 1950, p. 147.

Lucky, Robert W.: "A Survey of the Communication Theory Literature 1968–73," *IEEE Transactions on Information Theory,* vol. IT-19, 1973, pp. 725–739.

Schwartz, M.: *Information Transmission, Modulation and Noise,* McGraw-Hill, New York, 1970.

## Background Reading

Bennett, W. R., and J. R. Davey: *Data Transmission,* McGraw-Hill, New York, 1965.

Black, H. S.: *Modulation Theory,* Van Nostrand, New York, 1953.

Saltzberg, B. R.: "Intersymbol Interference Error Bounds with Application to Ideal Band Limited Signalling," *IEEE Transactions on Information Theory,* vol. IT-14, 1968, pp. 563–568.

Schwartz, M., W. R. Bennett, and S. Stein: *Communication Systems and Techniques,* McGraw-Hill, New York, 1966.

Taub, H., and D. Schilling: *Principles of Communication,* McGraw-Hill, New York, 1971.

Wozencraft, J. M., and I. M. Jacobes: *Principles of Communication Engineering,* Wiley, New York, 1965.

## PULSE CODE MODULATION (SAMPLING, QUANTIZATION, CODING)

### Supplementary Reading

Cattermole, K. W.: *Principles of Pulse Code Modulation,* Iliffe, London, 1969.

## Specialized Papers (Highly Recommended)

Bennett, W. R.: "The Spectra of Quantized Signals," *Bell System Technical Journal,* vol. 27, 1948, p. 446.

Davis, C. G.: "An Experimental Pulse Code Modulation System for Short Haul Trunks," *Bell System Technical Journal,* vol. 41, 1962, pp. 1–24.

Nyquist, H.: "Certain Topics in Telegraph Transmission Theory," *Journal of the American Institute of Electrical Engineers,* vol. 47, 1928, pp. 214–216.

Reeves, A.: "The Past, Present, and Future of PCM," *IEEE Spectrum,* vol. 2, 1965, pp. 56–63.

Shannon, C. E.: "A Mathematical Theory of Communication," *Bell System Technical Journal,* vol. 27, 1948, pp. 379–423, 623–656.

————: "Communication in the Presence of Noise," *Proceedings of the Institute of Radio Engineers,* vol. 37, 1949, pp. 10–21.

Smith, B.: "Instantaneous Companding of Quantized Signals," *Bell System Technical Journal,* vol. 36, 1957, pp. 653–709.

## Background Reading

Licklider, J. C. R., and I. Pollack: "Effects of Differentiation, Integration and Infinite Peak Clipping upon Intelligibility of Speech," *Journal of the Acoustical Society of America,* vol. 20, 1948, pp. 42–51.

Bennet, G. H.: *Pulse Code Modulation and Digital Transmission,* Marconi Instruments, Saint Albans, Hertfordshire, England, 1976.

Gibby, R. A., and J. W. Smith: "Some Extensions of Nyquist's Telegraph Theory," *Bell System Technical Journal,* vol. 44, 1965, pp. 1487–1510.

Mann, H., H. M. Straube, and C. P. Villars: "A Companded Coder System for an Experimental PCM Terminal," *Bell System Technical Journal,* vol. 41, 1962, pp. 173–226.

Shennum, R. H., and J. R. Gray: "Performance Limitations of a Practical PCM Terminal," *Bell System Technical Journal,* vol. 41, 1962, pp. 143–171.

Sugiyama, H.: "Band Limited Signals, and the Sampling Theorem," *Electronics and Communications in Japan,* vol. 49, 1966, pp. 100–108.

## MULTIPLEXING (SYNCHRONOUS PULSE JUSTIFICATION)

### Specialized Papers (Highly Recommended)

Brugia, O., and M. Decina: "Reframing Statistics of PCM Multiplex Transmission," *Electronics Letters,* vol. 5, no. 24, 1969, p. 623.

Haberle, H.: "Frame Synchronization PCM Systems," *ITT Electrical Communications,* vol. 44, 1969, p. 280.

## Background Reading

International Telegraph and Telephone Consultative Committee: *CCITT Orange Book,* vol. III-2, G733 (United States system), G732 (European system), International Telecommunications Union, Geneva, Switzerland, 1977.

Schwartz, L.: "Statistical Distributions of Frame Resynchronization Times in D1 and D2 type PCM Terminals," in *Proceedings of the Institute of Electrical and Electronics Engineers 1972 International Communications Conference,* Institute of Electrical and Electronics Engineers, Philadelphia, 1972, p. 21.

Vogel, E., and R. McLintock: "30 Channel PCM System: Part 1, Multiplex Equipment," *Post Office Electrical Engineers' Journal,* vol. 71, 1978, p. 5.

## MULTIPLEXING (ASYNCHRONOUS PULSE JUSTIFICATION)

### Specialized Papers (Highly Recommended)

Decina, H.: "Planning a Digital System Hierarchy," *IEEE Transactions on Communications,* vol. 20, 1972, p. 60.

Duttweiler, D. L.: "Waiting Time Jitter," *Bell System Technical Journal,* vol. 51, 1972, pp. 165–207.

International Telegraph and Telephone Consultative Committee: *CCITT Orange Book,* vol. III-2, G742 (recommendation), International Telecommunications Union, Geneva, Switzerland, 1977.

## Background Reading

Geissler, H.: *Planning a PCM Hierarchy,* NTZ Report 8, VDE-Verlag Gmbh, Berlin, 1971.


## DIGITAL TRANSMISSION (CODING, CLOCK EXTRACTION, JITTER, EQUALIZATION)

## Supplementary Reading

Bylanski, P., and D. G. W. Ingram: "Digital Transmission Systems," *Institution of Electrical Engineers Telecommunication Series 4,* Peregrinus, Stevenage, England, 1976.

## Specialized Papers (Highly Recommended)

Bellato, L., A. Tavella, and G. Vannucchi: *New Digital Systems over Physical Lines,* Telettra SPA., Milano, Italy.

Bennett, W. R.: "Statistics of Regenerative Digital Transmission," *Bell System Technical Journal,* vol. 37, 1958, p. 1501.

Byrne, C. J., B. J. Karafin, and D. B. Robinson: "Systematic Jitter in a Chain of Digital Regenerators," *Bell System Technical Journal,* vol. 42, 1963, p. 2679.

Cock, C. C.: "The Accumulation of Jitter in Digital Line Links," *Institution of Electrical Engineers Colloquium on Jitter in Digital Communication Systems,* London, England, 19 October 1977, digest no. 1977/45.

————: "Jitter Tolerances in Digital Equipments," *Institution of Electrical Engineers Colloquium on Jitter in Digital Communication Systems,* London, England, 19 October 1977, digest no. 1977/45.

Lender, A.: "The Duobinary Technique for High-Speed Data Transmission," *IEEE Transactions on Communications and Electronics,* vol. 82, 1963, pp. 214–218.

————: "Correlative Level Coding for Binary Data Transmission," *IEEE Spectrum,* vol. 3, 1966, pp. 104–115.

Manley, J. M.: "The Generation and Accumulation of Timing Noise in PCM Systems, an Experimental and Theoretical Study," *Bell System Technical Journal,* vol. 48, 1969, p. 541.

Mayo, J. S.: "Bipolar Repeater for Pulse Code Modulation Signals," *Bell System Technical Journal,* vol. 41, 1962, pp. 25–98.

## Background Reading

Bedrosian, E.: "Spectrum Conservation by Efficient Channel Utilization," *IEEE Communications Society Magazine,* vol. 15, 1977, pp. 20–27.

Bennett, G. H.: "Testing Techniques for 24 Channel PCM Systems," *Post Office Electrical Engineers' Journal,* vol. 65, 1972, p. 182.

Ericson, T., and H. Johansson: "Digital Transmission over Coaxial Cables," *Ericsson Technics,* vol. 27, 1971, pp. 191–272.

Franaszek, P. A.: "Sequence State Coding for Digital Transmission," *Bell System Technical Journal,* vol. 47, 1968, p. 143.

Fultz, K., and D. B. Penick: "The T1 Carrier Systems," *Bell System Technical Journal,* vol. 44, 1965, pp. 1405–1451.

Gerst, I., and J. Diamond: "The Elimination of Intersymbol Interference by Input Signal Shaping," *Proceedings of the Institute of Radio Engineers,* vol. 49, 1961, pp. 1115–1201.

Hirsch, D.: "A Simple Adaptive Equalizer for Efficient Data Transmission," *IEEE Wescon Technical Papers,* vol. 13, part 4, session 11/2 1969, 10 pp.

Kreztmer, E.: "Generalization of a Technique for Binary Data Communication," *IEEE Transactions on Communication Technology,* vol. COM-14, 1966, pp. 67–68.

Lucky, R. W.: "Automatic Equalization for Digital Communication," *Bell System Technical Journal,* vol. 44, 1965, pp. 547–588.

Pasupathy, S.: "Correlative Coding: A Bandwidth Efficient Signalling Scheme," *IEEE Communications Society Magazine,* vol. 15, 1977, pp. 4–11.

Richards, D. L., "Transmission Performance of Telephone Networks Containing PCM Links," *Proceedings of the Institution of Electrical Engineers* (London), vol. 115, 1968, pp. 1245–1258.

Smith, E., and M. Gabriel: "24 Channel PCM Junction Carrier System," *ITT Electrical Communications,* vol. 43, 1968, p. 123.

Waddington, D. E.: "PCM Link Maintenance," *Marconi Instrumentation,* vol. 11(3), 1967, p. 2.

Whetter, J., and N. Richman: "30 Channel PCM System, Part 2: 2.048 Mbit/s Digital Line System," *Post Office Electrical Engineers' Journal,* vol. 71, 1978, p. 82.

## DIGITAL SWITCHING

### Specialized Papers (Highly Recommended)

Perucca, Giovanni: "An Experimental Digital Switching System," *Proceedings of the International Switching Symposium,* Munich, Germany, Sept. 9–13, 1974, paper 227.

Pitroda, S.: "Selection of an Optimum Digital PCM Switching Configuration Based on a Set of System Considerations," *IEEE Conference Record of the International Conference on Communications, ICC 1974.*

————: "A Review of Telecommunications Switching Concepts," parts 1 and 2, *Telecommunications,* vol. 10, nos. 2 and 3, Feb. 1976, pp. 29–36, Mar. 1976, pp. 24–26, 28, 30.

### Background Reading

Altehage, G. A., and R. Slaban: "PCM Switching System EWSD," *Proceedings of the International Switching Symposium,* Kyoto, Japan, Oct. 25–29, 1976.

Bourkao, M., and J. Jacob: "New Developments in EIO Digital Switching Systems," *Proceedings of the International Switching Symposium,* Kyoto, Japan, Oct. 25–29, 1976.

Feiner, A.: "Switching Network ESS no. 1," *Bell Laboratories Record,* vol. 43, 1965, pp. 236–240.

McDonald, J. C., and J. R. Baichtal: "A New Integrated Switching System," *Proceedings of the National Telecommunications Conference,* Dallas, Texas, Nov. 29–Dec. 1, 1976.

Skaperda, N. J.: "Generic Digital Switching System," *Proceedings of the International Switching Symposium,* Kyoto, Japan, Oct. 25–29, 1976, vol. 1, session 223, pp. 4.1–4.8.

Smith, G. J.: "A Bid for the Digital Future," *Telephony,* July 19, 1976.

Vaughan, H. Earle: "An Introduction to no. 4 ESS," *Proceedings of the International Switching Symposium,* Cambridge, Massachusetts, June 6–9, 1972.

## FIBER OPTIC TRANSMISSION

### Specialized Papers (Highly Recommended)

Clarricoats, P. J. B. (ed.): *Optical Fibre Waveguides,* Peregrinus, Stevenage, England, 1975.

Kao, C., and G. Hockham: "Dielectric Fibre Surface Waveguides for Optical Frequencies," *Proceedings of the Institution of Electrical Engineers* (London), vol. 113(7), 1966, pp. 1151–1158.

Kapron, F. P., D. B. Keck, and R. D. Maurer: "Radiation Losses in Glass Optical Waveguides," *Applied Physics Letters,* vol. 17, 1970, pp. 423–425.

Ramsay, M. M., A. W. Horsely, and R. E. Epworth: "Subsystems for Optical-Fiber-Communication Field Demonstrations," *Proceedings of the Institution of Electrical Engineers* (London), vol. 123, June 1976, pp. 633–642.

### Background Reading

Berry, R., and R. Hooper: "Practical Design Requirements for Optical Fibre Transmission Systems," *Proceedings of the 1st European Conference on Optical Fibre Communication,* London, 16–18 September 1975, Institution of Electrical Engineers (London) Conference Proceedings no. 132.

Chown, M.: "Repeaters for Optical Communications Systems," *ITT Electrical Communications,* vol. 52, 1977.

Heinlein, W., and H. Trimmel: "Repeater Spacings of 8 Mbit/s, and 34 Mbit/s Transmission Systems using Multi-Mode Optical Waveguides and LED's," *Proceedings of the 1st European Conference on Optical Fibre Communication,* London, 16–18 September 1975, Institution of Electrical Engineers (London) Conference Proceedings no. 132.

Ramsay, M. M.: "The Use of Optical Fibre Waveguides in Communications Systems," *Communications International,* February 1977, pp. 52, 57, 58.

Randall, E., and R. Cerny: "Selection Considerations For Fibre Optic Cables," *Communications International,* February 1977, pp. 60–61.

# Index

# Index

# About the Author

FRANK F. E. OWEN is a physics graduate of Manchester University in England. He gained experience in telecommunications working for the International Telephone and Telegraph Company, first at their research laboratories in Harlow, England, and later at the product development center in Milano, Italy. He is presently with Texas Instruments European Headquarters in Nice, France, where he is responsible for telecommunications product development and marketing strategy.

Prior to joining Texas Instruments, Mr. Owen served for three years as an advisor on digital telecommunications to the Brazilian government under the U.N. Development Program. In this capacity, he advised on PCM multiplexer and transmission equipment development later to be manufactured for Telebras, the Brazilian national telecommuncations company.